Real-Time Three-Dimensional Imaging of Dielectric Bodies Using Microwave/ Millimeter-Wave Holography

Real-Time Three-Dimensional Imaging of Dielectric Bodies Using Microwave/ Millimeter-Wave Holography

Reza K. Amineh, Natalia K. Nikolova, and Maryam Ravan

IEEE PRESS

WILEY

Published by John Wiley & Sons, Inc., Hoboken, New Jersey.
Published simultaneously in Canada.

For general information on our other products and services or for technical support, please contact our Customer Care Department within the United States at (800) 762-2974, outside the United States at (317) 572-3993 or fax (317) 572-4002.

Wiley also publishes its books in a variety of electronic formats. Some content that appears in print may not be available in electronic formats. For more information about Wiley products, visit our web site at www.wiley.com.

Library of Congress Cataloging-in-Publication Data is available.

Paperback ISBN: 9781119538868

Printed in the United States of America.

V10012170_071519

To Rose

Contents

Preface

This book reviews microwave and millimeter-wave imaging techniques based on holographic principles. These techniques are used by the whole-body imagers in the airports and other public places as part of routine security screening. Such devices exemplify the importance and practicality of the holographic image reconstruction techniques within a variety of other microwave imaging methodologies, most of which have been implemented in a controlled lab environments only. The book provides a condensed review targeting a broad audience including faculty, graduate students, and researchers working in the general area of microwave and millimeter-wave imaging and sensing. The outline of the book is as follows:

In Chapter 1, a summary of the microwave imaging techniques and their applications is provided with a focus on the holographic imaging techniques.

Chapter 2 starts with the primary holographic imaging concepts developed in optics and then describes how these concepts were extended to microwave and millimeter-wave imaging.

In Chapter 3, recent developments of wide-band microwave holography with depth resolution are discussed. These developments are derived from holography techniques that utilize phase and amplitude information recorded over a two-dimensional aperture to reconstruct a focused image of the object using computer algorithms emulating Fourier-optics-based image reconstruction. The evolution of these techniques from the synthetic aperture radar principles is also covered. Development of rectangular and cylindrical setups will be discussed together with their capabilities and their limitations.

Chapter 4 discusses the adaptation of these techniques to near-field applications such as those arising in biomedical imaging, nondestructive testing, and underground imaging. These techniques utilize phase and amplitude information recorded over a two-dimensional aperture and perform digital image reconstruction. However, they do not impose far-field approximations on the waves (plane or spherical waves). Two recently developed approaches are utilized: (i) extracting near-zone incident-field and Green's function distributions, and (ii) using the measured point-spread function of the imaging system.

Chapter 5 starts with introducing the well-known "diffraction limit" in the resolution. Also discussed are some techniques to overcome this limit are discussed such as the use of super-oscillatory filters and the use of resonant scatterers in the near-field of the imaged objects. The chapter then discusses methods of obtaining quantitative images of the inspected medium.

In Chapter 6, the capabilities and limitations of the holographic reconstruction techniques discussed in the book are summarized. Also, recommendations are made to overcome some of these limitations and to expand the applications of the microwave/millimeter-wave holographic techniques.

Reza K. Amineh
Natalia K. Nikolova
Maryam Ravan

Acknowledgments

We are indebted to numerous colleagues and students who helped us develop the near-field holographic imaging techniques. This book would not exist without their effort and we would like to acknowledge these people here: Prof. George Eleftheriades at the University of Toronto and a few graduate students at McMaster University including (in alphabetical order) Yona Baskharoun, Ali Khalatpour, Justin McCombe, Kaveh Moussakhani, Denys Shumakov, Daniel Tajik, Aastha Trehan, Sheng Tu, and Yifan Zhang. We would also like to thank Wiley-IEEE Press for their encouragement and help while writing the book. Our deepest gratitude goes to our parents for their unwavering support during the many years of education and professional growth.

1

Introduction

Microwave (300 MHz to 30 GHz) and millimeter-wave (30–300 GHz) imaging (MMI) technologies exploit electromagnetic (EM) waves with much lower frequencies than the visible light. This allows for sufficient penetration into various dielectric materials for inspection purposes and the ability to "see" through media which cannot be inspected via optical or acoustic means. These technologies have underwent rapid growth during the last few decades due to a steady demand for new applications and services along with remarkable miniaturization and diversification of the high-frequency electronics. This has resonated well with the significant developments in computing technologies and processing algorithms leading to a wide range of applications such as imaging various composites, cement, soil, wood, ceramics, plastics, clothing, living tissues, etc.

Another major motivation to develop these technologies is that the microwave and millimeter-wave radiation is nonionizing and is not considered hazardous at moderate power levels. This makes such technologies favorable compared to the competing X-ray imaging technology which is ionizing, i.e. it has enough energy to potentially cause cellular and DNA damage or to elevate the risk of developing cancer in living tissues [1].

Moreover, the possibility of measuring wideband data as well as acquiring both phase and magnitude (which is typically challenging at frequencies higher than microwave and millimeter-wave) makes these technologies favorable for three-dimensional (3D) imaging applications, e.g. see [2]. All the aforementioned advantages have motivated the researchers in academic and industrial institutions to pursue development of the microwave technologies toward various imaging and sensing applications.

Figure 1.1 illustrates a typical MMI system consisting of a transmitting antenna illuminating the inspected medium. The inspected medium includes the background along with any objects that are electrically sufficiently different

Real-Time Three-Dimensional Imaging of Dielectric Bodies Using Microwave/Millimeter-Wave Holography, First Edition. Reza K. Amineh, Natalia K. Nikolova, and Maryam Ravan.
© 2019 by The Institute of Electrical and Electronics Engineers, Inc.
Published 2019 by John Wiley & Sons, Inc.

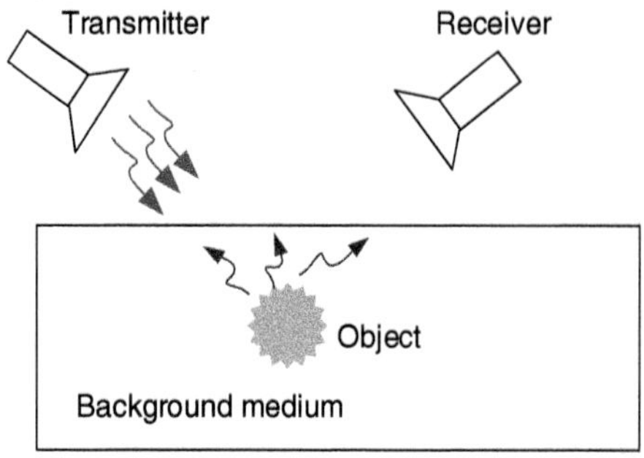

Figure 1.1 Illustration of a MMI setup. The object inside the background medium scatters the EM waves emitted by the transmitter provided its electrical properties are different from those of the background. Some of the scattered waves are picked up by the receiving antenna and are processed to detect the object.

from the background to cause scattering of the EM waves. The scattered waves due to the presence of the object along with the scattered waves due to the background medium are acquired by a receiving antenna. The acquired data are then processed to reconstruct two-dimensional (2D) or 3D images of the object.

In various MMI systems, the transmission and reception can be implemented by the same antenna (monostatic system), or one antenna can transmit while another receives (bistatic system), or an array of antennas (multistatic) can receive the back-scattered waves. Besides, the data acquisition can be implemented while the antennas are stationary or the antennas may perform scanning of the back-scattered waves over an aperture to construct a synthetic aperture. An alternative solution to avoid mechanical scanning of the antennas (that may be time-consuming and may suffer from positioning errors) is the use of stationary antenna arrays that can be switched electronically to capture the distribution of the back-scattered waves over an aperture.

1.1 Some Emerging Applications of MMI

Some of the applications of microwave imaging technology, such as synthetic aperture radar (SAR) [3] and ground penetrating radar (GPR) [4], have been well-established, commercialized, and employed for several decades. These technologies provide high-resolution qualitative images of the earth surface, landscape, weather condition, underground objects, hydrocarbon reservoirs, etc. These have been typically long-range applications of MMI since the distance between the imaged objects and the antennas (sensors) is typically much larger than the operation wavelength.

Recently, researchers have been pursuing the capabilities of MMI in some new applications, demonstrating promising imaging results. These new applications, such as concealed weapon detection [5], nondestructive testing (NDT) [6], through-the-wall imaging [7], and biomedical imaging [8], are under rapid development and commercialization. Their success depends on how they can overcome challenges encountered by the competing technologies of ultrasound imaging, X-ray imaging, magnetic resonance imaging (MRI), etc. in each particular application domain.

The conventional MMI techniques such as SAR and GPR suffer from the well-known "diffraction limit" in the resolution. The diffraction limit states that the spatial resolution of the far-field techniques is proportional to the operation wavelength, i.e. the shorter the wavelength, the smaller the size of the shapes that can be resolved in the image. However, near-field MMI techniques may offer better spatial resolution which is typically proportional to the sensor's sensing dimensions (e.g. see [9]). This is due to the use of evanescent waves, which contain information about finer details of the object but decay exponentially versus the distance between the object and the antenna. Thus, while using the evanescent-wave information is possible in the near-field MMI applications (leading to better spatial resolution), in the far-field, receiving the evanescent waves is practically impossible leading to degradation of the spatial resolution down to the diffraction limit. Later in this book, we discuss the resolution of MMI systems in more detail.

The case of interest in the recent applications of MMI is that of an object in the near field of the antenna. This necessitates the development of new algorithms and techniques for image-reconstruction purposes. In the following, we briefly describe the recent progress in some of the emerging and most promising applications of MMI technologies which are mostly mid-field to near-field applications.

One revolutionizing application of the MMI is in security screening in public places such as airports. This technology allows for penetration of the EM waves through clothes and for the formation of a whole-body image of the scanned person in order to seek for prohibited concealed items [5, 10–13]. So far, several countries have chosen scanners based on this technology for security screening in the airports [14] over the competing technology, which is back-scattered X-ray. These millimeter-wave scanners can be divided into two categories: active and passive. Active scanners illuminate the inspected domain with millimeter-wave energy and then measure and process the reflected energy for image reconstruction [5, 11–13]. Passive systems create images using only the millimeter-wave spectrum of the ambient thermal radiation and that emitted from the inspected body [15–17].

Figure 1.2 shows a millimeter-wave imaging setup used by transportation security administration (TSA) in the US airports for whole-body scanning. This figure also shows some sample reconstructed images by this system. It is worth noting that the resolution of the produced images using millimeter-wave technology in the airport security screening is so high that it has raised privacy concerns. This has led to the prohibition of displaying detailed images by the US congress and applying automatic target recognition software (which provides a generic body outline with largest contrasts due to the concealed prohibited

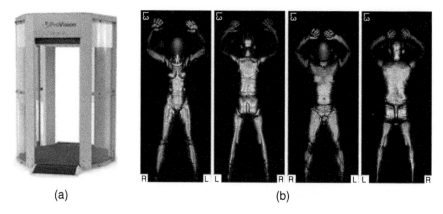

(a)   (b)

Figure 1.2 (a) Millimeter-wave whole body scanning setup based on the technology developed for airport security screening. (b) Sample images of screened persons. *Source:* Reprinted with permission from [14].

objects). This is to emphasize that very high-resolution, robust, and accurate images can be captured using this technology.

Another application of the MMI, which is under active research in the microwave society, is in biomedical imaging [18]. This includes imaging of isolated organs or *in situ* imaging such as imaging of the brain for stroke detection [19–21], cerebral edema [22], breast cancer [8, 23–25], bone imaging [26, 27], heart imaging [28–33], and joint-tissue imaging [34]. In these applications, the tissue is illuminated with low-power microwaves which are safe and allow for frequent examinations. The scattered waves are then measured and processed to provide images of the interior of the tissue. The images show various tissues such as muscle, fibro-glandular and fatty tissues, which exhibit various contrasts in the dielectric properties. This provides a basis for distinguishing the normal tissue from the malignant tissue (tumor). The malignant tissue typically has larger dielectric properties compared to normal tissue due to the higher concentration of blood vessels and water content. Figure 1.3 shows the measurement setup for brain imaging using a physical head phantom and the Vivaldi antenna proposed in [35]. This figure also shows the reconstructed images using the confocal method showing damage due to stroke at two different locations.

Biomedical microwave imaging is proceeding fast to make an entrance in the highly competitive world of medical imaging. Recently, there have been start-up companies that are commercializing the microwave tissue imaging technology such as EMTensor [36] and Medfield Diagnostics [37] for stroke diagnosis, and Micrima [38] for breast imaging.

In general, the biomedical imaging technology based on MMI offers advantages such as low health risks, ease of mobility due to compactness, relatively good resolution, and cost effectiveness. However, it often lacks the spatial

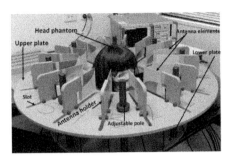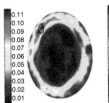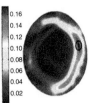

Figure 1.3 (a) Measurement setup for brain imaging using a physical head phantom and the Vivaldi antenna. (b) Reconstructed images using the confocal method showing position of stroke at two different locations. The true position of the stroke is shown by an ellipse. *Source:* Reprinted with permission from [35].

resolution that may be required for some applications. Presently, its spatial resolution is inferior compared to that of computed tomography (CT) or MRI. The success of MMI technologies in the biomedical field depends on the capability to address the technical and commercial challenges faced by the competing technologies. One major challenge in this area, which is currently under intensive investigation, is the relatively high error rate that can occur when the electric contrast between malignant and healthy tissues is not sufficiently large, as in the case of tumors in fibro-glandular tissue or muscle [39, 40]. This dramatically reduces the specificity of a microwave-based diagnosis. Arguably, this may be the main challenge in applications such as breast cancer screening, since most of the breast tumors are located in this kind of tissue. To overcome this problem, the use of contrast agents such as carbon nanotubes [41] and magnetic nanoparticles [42–46] is gaining attention. In particular, the use of magnetic nanoparticles has been under investigation to take advantage of their specific features compared to dielectric contrast agents [42]. These particles have been approved for clinical use and can be manipulated properly so that they can specifically target the malignant tissues. One important feature is that their response in the microwave regime can be properly changed by applying an external polarizing magnetic field. This external field does not change the response of the normal tissue which has not been targeted by these particles. As a result, by exploiting a proper differential measurement strategy, one can separate the response of the normal tissue from that of the malignant tissue targeted by the magnetic nanoparticles [42]. The feasibility of this technique has been investigated and has demonstrated promising results in simulations and preliminary experiments.

The last important application of MMI that is discussed here is the NDT of nonmetallic media for integrity inspections. NDT based on MMI involves developing sensors/probes, methods, and calibration techniques for detection of flaws, cracks, defects, pits, voids, inhomogeneities, moisture content (MC), hidden artifacts, etc. by means of microwaves and millimeter waves [6, 47]. Such technology is increasingly being used for quality control and condition assessment of various dielectric materials with the following diverse objectives: detection and sizing of fatigue cracks in metal surfaces including those hidden under paint and dielectric coatings [48], detection and evaluation of corrosion under paint and composite laminates [49], defect evaluation in various composite structures [50], evaluation of the composites [6, 47], and microwave microscopy [51].

A major emerging application of MMI is the inspection of composite materials. These materials are replacing many industrial metallic components due to being lighter, stiffer, stronger, and more durable. Composite materials cannot be inspected by the conventional NDT techniques such as eddy current or magnetic flux-leakage measurements which can only be employed for inspection of metallic structures. On the other hand, MMI techniques are well-suited for testing these materials since microwaves and millimeter waves can easily penetrate low-loss dielectric media. In fact, in many cases, MMI could be the only technology that can be employed for integrity inspections. In some other cases, it could be combined with other NDT techniques such as ultrasound imaging to provide a more comprehensive inspection. A good review of microwave and millimeter-wave NDT techniques has been provided in [52]. Figure 1.4

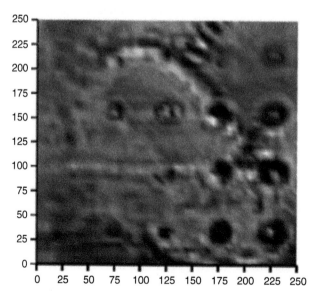

Figure 1.4 Sample millimeter-wave image of a shuttle external fuel tank's spray-on-foam insulation (SOFI) at 100 GHz using a horn antenna [50].

shows a sample millimeter-wave image of a shuttle external fuel tank's spray-on-foam insulation (SOFI) at 100 GHz using a horn antenna [50].

In general, MMI offers several advantages over other NDT methods such as radiography, ultrasonics, and eddy currents [52]: relatively low cost, good penetration in dielectric materials, good resolution especially in near-field applications, contactless feature of the microwave sensor (antenna), no need for a couplant to transmit the signal into the material under test (unlike ultrasonic methods), monostatic, relatively low power, compact, easily adaptable to existing industrial scanners, real time, no need for operator expertise in the field of microwave engineering, robust, rugged, and repeatable. Given the outlook for the rapid and steady growth of incorporating composite materials in a wide range of structures (replacing metallic components), the future of microwave and millimeter-wave NDT is very promising.

1.2 Quantitative Versus Qualitative MMI

In terms of the accuracy of the reconstructed images, MMI techniques can be classified as either quantitative or qualitative. These techniques vary significantly with regard to the hardware components, data acquisition schemes, and processing algorithms. The reader is referred to [53] for a comprehensive review of various MMI techniques.

In quantitative imaging techniques, which are also commonly referred to as inverse scattering methods or optimization-based methods (e.g. see [54]), the aim is to provide an accurate estimation of the electrical properties of the imaged objects. This is accomplished by solving a nonlinear inverse problem. Such solution is typically expensive to evaluate in terms of both memory and time. Thus, although these techniques provide more accurate images of the inspected media, they are not practical for real-time applications such as object tracking or tissue imaging in which the fast motion of the object or patient, respectively, may create artifacts. In quantitative imaging techniques, the nonlinear inverse problem can be converted into a linear inverse problem by using the Born or Rytov approximations [55]. Despite the fact that direct matrix inversion methods can be employed to solve the reconstruction problem, this can be costly when the size of the problem (number of voxels in the inspected domain) is large. To overcome this problem, direct inversion of the relevant matrix is replaced with iterative solvers [56–64]. These techniques are also usually slow and not practical for real-time imaging applications.

It is worth noting that recently two direct quantitative imaging approaches have been proposed [65]. One is based on the scattered-power maps [53, 66, 67]. The second one exploits the principles of microwave holography [68]. The accuracy and the applicability of these methods are limited due to the fact that, as any other linear-inversion method, these methods are based on a linearized model of scattering. Their advantage is that they are fast and capable of providing images in real time. Both methods rely on the use of quantitatively accurate point-spread function (PSF), i.e. the PSFs must not only represent the normalized magnitude and phase distribution of the response to a point scatterer but also they have to scale properly, in a complex sense, with the complex contrast of the point scatterer. Quantitatively accurate PSFs can be acquired via calibration measurements. The closer the background medium in these measurements is to the averaged electrical properties of the imaged object, the better the quantitative output would be since the method remains rooted in a linearized model of scattering. In far-zone measurement scenarios, "normalized" PSF approximations such as the plane-wave and the isotropic-wave approximations could also be used. These can be additionally improved if the antenna radiation pattern is employed to correct the field distribution. Such approximations eliminate the need to perform calibration measurements and would, therefore, save time.

In theory, the advantage of quantitative optimization-based techniques is that they are capable of accurate estimation of the contrast function of the imaged object. However, in practice, these techniques suffer from serious challenges such as complexity, prohibitive memory, and time requirements, as well as ill-posedness of the solution. In particular, the solution is typically sought in an iterative manner in which at each iteration a 3D forward EM problem needs to be solved. This is very time-consuming and makes the use of these techniques in real-time imaging applications almost impossible (considering the current limitations of the computing technology). In contrast, most qualitative techniques often provide less accurate images but are much faster. Typically, by applying proper simplifying assumptions on the scattering model, these techniques seek an approximate solution to the inverse scattering problem that can be as fast as a fraction of a second even with tens of thousands of voxel contrast values. In these techniques, a qualitative profile (which is referred to as reflectivity function, qualitative image, or contrast function) is reconstructed. The aim is to estimate the position, the shape, and the spatial extent of an object whereas its contrast with respect to the background medium may be estimated only approximately, often in the form of a normalized image of the reflectivity.

The qualitative MMI techniques can be categorized into time-domain imaging (e.g. see [69]), diffraction tomography (DT) (e.g. see [70]), sensitivity-based imaging [71], and holography (e.g. see [72]). In general, most of these techniques use radar concepts to generate a qualitative reflectivity image in order to differentiate the object from the background. The objects are usually stronger scatterers than the background medium. In the following, we briefly review these techniques.

The time-domain imaging systems operate based on the radar signal-processing principles in which the inspected medium is illuminated with wide-band pulses and the back-scattered pulses due to the presence of the object are measured. Various data collection schemes have been employed according to the specifics of the application such as stationary antennas or antennas that scan over an aperture to construct a synthetic aperture. The acquisition technique where the radar collects data over a line or a surface is referred to as SAR. Nowadays, time-domain imaging systems use ultra wideband (UWB) pulses, i.e. the pulses may cover a wide frequency band all the way from 100 MHz to about 10 GHz. The UWB radar is a rapidly growing technology which holds enormous potential for further applications in detection, imaging, and guidance [73, 74].

DT emerged in the late seventies and the eighties mainly in the fields of acoustics and microwave imaging as an alternative to the straight-ray tomography developed for X-ray imaging. Its principles are closely related to those of the holographic reconstruction. It could be considered as a generalization of the conventional straight-ray tomography used in X-ray CT. Unlike X-rays, acoustic and EM waves do not travel along straight paths. This is because the components of the imaged objects are often comparable or smaller than the wavelength. The inversion in X-ray tomography is based on the Fourier Slice Theorem, which assumes a straight-path propagation model. Generally, such model cannot be applied in acoustics or microwave imaging. Instead, when the wavelength is long, an alternative method called Fourier Diffraction Theorem is employed to reconstruct a 3D image of the object [75]. It can be shown that the Fourier Diffraction Theorem reduces to the Fourier Slice Theorem when the wavelength tends to zero [75]. The Fourier Diffraction Theorem employs the linearized scalar Born model of scattering. It also assumes a plane-wave illumination and point-like receiver antennas. Plane-wave illumination of the whole imaged region is not easy to implement at microwave frequencies, especially in the near-field imaging problems. Thus, the need to place the object in the far zone of the microwave source is a serious limitation in DT. Due to this reason, DT is not convenient for near-field imaging applications such as tissue imaging [76–78]. Besides, to implement DT, the receiver antennas must closely resemble point-wise omnidirectional probes. This is also not practical in near-field imaging. Due to such limitations, DT has not made much progress in terms of real-world applications, in particular, for near-field imaging of dielectric bodies.

Another qualitative MMI technique is sensitivity-based imaging that was first proposed in [79, 80] as a detection method. Its spatial resolution and robustness to noise was later studied in [71]. It employs frequency-domain responses and it can be configured to work with complex-valued responses, or with magnitude and phase responses separately, depending on which is available. In far zone reflection (back-scattered) measurements, it is the phase that carries most of the information about the object and its availability is critically important. In the extreme near zone measurements, back-scattered or forward-scattered, magnitude-only data may suffice. A far zone scenario where magnitude-only data may be sufficient arises in the imaging of lossy objects where the signal attenuation is the decisive factor shaping the signal. This imaging technique has been extended to quantitative imaging as described earlier.

The last qualitative MMI technique described here is holographic imaging which dates back to the early 1970s when the original optical holography developed in [81, 82] was extended to longer wavelengths at microwave frequencies [83–85]. Almost at the same time, similar approaches were developed for acoustic holography. In all three areas (optics, acoustics, and microwaves) the amplitude and phase of a wavefront scattered from the inspected object is captured first via recording a hologram. The sampled wavefront is then "reconstructed" either optically or by using Fourier-optics-based computer image-reconstruction algorithms. Acoustic and microwave holography are well suited to digital computer image reconstruction since the data are typically sampled by a scanned transceiver, and the relatively long wavelength allows for reasonable data size when fulfilling the Nyquist sampling criterion.

The modern holographic MMI techniques were originally proposed for concealed weapon detection. They used far-field approximations for the illuminating and back-scattered waves along with point-wise transmitter and receiver antennas. However, these techniques have been developed further in recent years to provide more accurate imaging results for near-field applications and for large transmitter and receiver antennas. These developments expand the range of applications for holographic MMI techniques significantly. For instance, they can be employed to realize real-time 3D imaging of the tissues with higher accuracy. Various data acquisition configurations have been developed for holographic MMI based on the rectangular apertures (e.g. see [5, 86, 87]), cylindrical apertures (e.g. see [11, 88, 89]), and manually selected and nonuniformly distributed measurement positions [90].

1.3 Advantages of Holographic MMI Techniques

For real-time 3D imaging of dielectric bodies, as described earlier, the quantitative imaging techniques are often not practical due to the huge memory and time requirements. While the qualitative techniques are faster, they suffer other drawbacks mostly with regard to the accuracy of the generated images. This is due to the simplifying assumptions made to expedite the solution of the inverse scattering problem. Some of these approximations include: (i) assuming point-wise transmitter and receiver antennas, (ii) assuming locally plane-wave form for the illuminating and back-scattered waves, and (iii) using the linear Born approximation which implies that the field inside the object is approximated with the incident field, i.e. the field that would be observed in the absence of the object.

Recently, holographic imaging techniques have been developed to address the first two issues. These techniques can now take into account the physical structure of the antenna and its effect on the produced incident field. These are of crucial importance when using MMI techniques for near-field imaging applications such as tissue imaging or NDT. In this book, we review the breakthrough developments in holographic MMI, which have paved the way toward practical real-time 3D imaging in both far-field and near-field applications.

1.4 Chronological Developments in the Holographic MMI Techniques

Holographic MMI techniques were originally rooted in the concepts derived in the optical regime but some of the later developments showed their close resemblance to the SAR imaging methods. Here, we describe how holographic MMI techniques emerged and evolved along a path toward similarity with SAR imaging and how they were further extended for near-field applications.

The holographic imaging concept was originally introduced in the field of optical imaging when Gabor [81, 91] aimed at improving the images obtained by electron microscope although he demonstrated the feasibility of his method with light waves. For this purpose, he demonstrated the possibility of acquiring the magnitude and phase of a wave as an interference pattern formed by this wave and a known reference wave. He also developed the mathematical foundations for the image reconstruction from such interference patterns. A review can be found in [92]. Gabor's experimental setup was improved by Leith and Upatnieks [82, 93] in the early sixties to achieve higher quality images. The work of Leith and Upatnieks resonated well with the availability of lasers leading to a new generation of 3D imaging systems. The progress in the field of optical holography inspired researchers in other fields, in particular acoustics [94–96], to use similar concepts for imaging optically opaque objects. In other words, while in the original optical holography the aim is to reproduce the optical perception of a 3D object from its recording, i.e. when the object is no longer available for viewing, in acoustic holography the objective is to produce an image of an object which is embedded inside an optically obscured medium. Despite this difference, the principles of producing an image are common between optical and acoustic holography.

Microwave holography was developed almost at the same time as acoustical holography [97–101]. Similar to acoustical holography, microwave holography aims at the image reconstruction of optically opaque objects. Thus, in this early stage, microwave holography closely resembled the optical method for data acquisition referred to as the "recording step" and that for the image reconstruction referred to as "reconstruction step" [102, 103]. In the recording step, an intensity pattern (hologram) is formed from the interference pattern between the scattered wavefront and a coherent plane-wave, introduced at an offset angle to the recording plane. This offset hologram was first introduced by Leith and Upatnieks [82], for the recording of optical wavefronts. In an optical hologram, the record is imprinted on a photographic plate (the recording medium). In microwave holography, this could be the intensity pattern acquired through the scanning of an antenna over the acquisition plane, the antenna being connected to a simple diode detector (a receiver that measures only the signal intensity) [85].

The subsequent progress in the field of holographic MMI can be divided into two separate paths: indirect holography and direct holography. What is common between these two categories of techniques is that both exploit the magnitude and the phase information of the back-scattered waves in order to generate images and to achieve resolution enhancement compared to the images obtained from raw amplitude measurements only.

The indirect microwave holography has been in fact the extension of the early work on this topic. It has the most resemblance to optical holography in terms of recording of a hologram. The major advantage of indirect holography is that it does not require the use of expensive vector-measurement equipment to obtain the complex field scattered from the imaged object. Instead, the phase information is mathematically recovered from low-cost intensity-only scalar microwave or millimeter-wave measurements. This simplifies the hardware implementation and reduces the cost of the imaging system. In the literature, indirect microwave holography was proved to have a significant potential to be employed in the measurement of complex antenna near-field and far-field radiation patterns [104, 105]. Recently, this technique has been extended to the imaging of early-stage breast cancer tumors [106] and metallic objects [107]. Although promising results have been achieved in [106, 107], the reconstruction of the scattered field was limited to the measurement plane. In [108], this issue has been addressed by back-propagating the reconstructed scattered field at the measurement plane to the plane of the imaged object.

The direct holographic MMI techniques, however, use radio frequency (RF) circuitry or vector network analyzer (VNA) to record both the magnitude and phase of the back-scattered waves directly. Although, similar to recording a hologram in indirect holography, measurements are typically performed over an aperture (plane), the scattered waves are sampled in complex form (unlike the recording of scalar values for capturing a hologram in indirect holography). As the direct holographic algorithms developed, their resemblance to SAR-based imaging methods became evident. Currently, the terms SAR imaging and direct microwave holography are being used to refer to reconstruction approaches that are mathematically very similar, in particular, for far-field imaging applications. In both, exploiting the phase information is at the heart of the image generation. The difference is usually in the dimensionality of the problem. SAR works with the wideband radar data acquired on a line. The location of the imaged surface is known. Thus, SAR methods reconstruct a 2D image of the object reflectivity on this known surface. In contrast, direct microwave holography works with wideband data acquired on a surface which enables the reconstruction of a 3D image. In particular, the proposed direct holographic technique in [5, 11] merges the single-frequency 2D holography with wideband 2D SAR to achieve 3D imaging. Wideband back-scattered signals are collected over a rectangular [5] or cylindrical [11] aperture. The processing relies on an assumed analytical (exponential) form of the incident field and the Green function to cast the inversion in the form of a 3D inverse Fourier transform (FT). An improvement of this imaging approach has been presented in [90] where the nonuniform manual scanning has been proposed as a practical imaging method.

In [86, 109, 110], the direct 3D holography technique was extended to near-field imaging. In this case, the analytical approximations of the incident field and the Green's function are inadequate; this is why numerical models are employed instead, which better represent the particular acquisition setup and antennas. The method proposed in [86, 109, 110] also allows for incorporating forward-scattered signals in addition to the back-scattered signals acquired on two opposite planar surfaces. The processing involves the solution of a linear system of equations for each spatial frequency pair (k_x, k_y). Then, 2D inverse FT is applied to the solution on planes (slices) at all desired range locations. The advantage of this approach is that the linear systems of equations have much smaller dimensions and are less ill-conditioned compared with the systems of equations arising in optimization-based imaging techniques. Furthermore, the resampling of the data in k_z-space is avoided. The assumption that k_x, k_y, and k_z are independent variables, which leads to errors in the image reconstruction process, is also unnecessary since the use of the parameter k_z is avoided altogether. It is important to note that in [109, 110], the role of the back-scattered data in 3D holographic imaging with planar scanning is emphasized, particularly with regard to the range spatial resolution. This is relevant in the imaging of lossy dielectric mediums such as tissues where the back-scattered data are of much poorer signal-to-noise ratio (SNR) compared with the forward-scattered data, resulting in loss of range resolution. In the near-field holographic imaging [86, 109, 110], the data for the incident field and the Green function are obtained via simulations. In practice, often, the fidelity of the simulation models, although better than the analytical approximations, may still be too low to ensure good image quality. The fidelity of the simulation model can be assessed through its ability to reproduce the measurements of known objects performed with the particular imaging setup. One factor contributing to the low fidelity of the simulations in near-field imaging can be the numerical errors such as those due to a coarse discretization mesh or imperfect absorbing boundary conditions. Such errors however, are not the major concern because they can be reduced by mesh refinement and stricter convergence criteria for the numerical solution. Unfortunately, this may also lead to prohibitive computational burden. The major concern in near-field imaging is in the so-called modeling errors which are much more difficult to reduce. These errors are rooted in the inability to predict all influencing factors arising in the practical implementation of the acquisition setup. These include: fabrication tolerances of the antennas and the positioning components, uncertainties in the constitutive parameters of the materials (especially the absorbers) used to build the measurement chamber, deformations due to temperature or humidity, and so on. In addition, models often ignore complexities in the cables, the connectors, the fine components of the measurement chamber (e.g. screws, brackets, thin supporting plates), and so on. In [87], a method has been proposed to acquire the incident field and the Green's function specific to the particular acquisition system via measurements of a known calibration object (CO). The method exploits the concept of PSF of a linear imaging system where the response due to an arbitrary object is the convolution of the response due to a point-wise scatterer (the scattering probe) with the spatial variation of that object.

1.5 Future Outlook for Holographic MMI for Real-Time 3D Imaging Applications

The authors of this book believe that direct holographic MMI provides a powerful and robust means for practical real-time 3D imaging and is projected to grow rapidly in the coming years. The main driving force behind this growth is both the demand for inspection and imaging of media for which MMI is the best method of choice. It is envisioned that this growth will be boosted even further due to the unprecedented miniaturization of the radio-frequency hardware as well as the recent progress in the processing algorithms for the near-field imaging applications.

On the hardware side, the size and the cost of the hardware needed to perform microwave sensing has dramatically decreased with the advent of the radio-on-a-chip (RoC), the on-chip software-defined radios (SDR), and the single-chip radars operating well into the 70-GHz bands. Coherent signal measurements with amplitude and phase information are now possible with electronic circuits occupying areas less than a square centimeter and with a price that ranges anywhere between tens and several hundreds of dollars. Multiplexing hundreds even thousands of microwave sensors through digital control is within reach allowing for bypassing the expensive and bulky RF switching and power-distribution networks. As a result of these advancements, large arrays of microwave sensors which are capable of sampling the back-scattered wave in complex form (acquisition of both magnitude and phase) are now affordable.

On the processing side, the availability of fast, robust, and accurate algorithms encourages the use of direct holographic MMI technology for many new applications, in particular, those for which the imaged object is in the near-field of the sensing antennas. To the best of our knowledge, there is no other MMI technique that can offer all the advantages of near-field direct holographic MMI techniques including: (i) being fast and robust, (ii) take into account the physical size of the antennas and their near-field, and (iii) taking advantage of the evanescent waves in the near-field of the imaged object to achieve the highest cross-range resolution.

2

Microwave/Millimeter Wave Holography Based on the Concepts of Optical Holography

Holographic imaging was first introduced in the field of optics in the late forties, when Gabor aimed at improving the quality of the images in electron microscopy [81, 91]. He proposed an approach to capture the magnitude and phase of a wave by means of an interference pattern formed by that wave and a known reference wave. He also developed the mathematical approach to the image reconstruction from such interference patterns. Later, in the early sixties, Leith and Upatnieks [82, 93] improved Gabor's experimental setup. In their method, the reference wave is a distinct beam separated from the illuminating beam by using a beam splitter. The reference beam travels at a substantial angle to the illuminating beam and is so inclined as to never illuminate the object. This new method of providing the reference wave led to separation of the overlapping images originally produced in Gabor's experimental setup in which the reference and object waves were nearly parallel. As a result of the work of Leith and Upatnieks, higher quality images were obtained. Since then the holography method has significantly influenced optics. In general, since optical recording media such as photographic films only respond to intensity, optical holography is applied as an indirect means of capturing the phase information of the optical wave fronts. It is implemented in two steps. The first step is the hologram formation step and the second step is the wave front reconstruction step. Later, we describe these two steps in more detail.

Soon after the introduction of holography in the field of optics, the imaging principles were expanded to other wave fields, most notably acoustics and microwaves. Acoustical holography is an active field with many similarities to microwave holography because the wavelengths employed in the two fields differ only by an order of magnitude. In general, at earlier stages, acoustical holography and microwave holography closely resembled the optical method for data acquisition (the hologram recording step) and that for the image reconstruction (the reconstruction step). But in more recent developments, digital computer image reconstruction is more common for acoustic and microwave

Real-Time Three-Dimensional Imaging of Dielectric Bodies Using Microwave/Millimeter-Wave Holography, First Edition. Reza K. Amineh, Natalia K. Nikolova, and Maryam Ravan.
© 2019 by The Institute of Electrical and Electronics Engineers, Inc.
Published 2019 by John Wiley & Sons, Inc.

holography. This is due to the fact that the hologram is typically sampled by scanning a transceiver and the relatively long wavelength (compared to optics) allows for reasonable data size if an attempt is made to fulfill the Nyquist sampling criteria.

In this chapter, we describe what is known in some literature sources as true microwave holography rather than quasi-microwave holography. In the quasi-microwave holography discussed in the next chapters, pulsed or multi-frequency radiation is utilized and the measurements are performed along a line (or a plane). Then two-dimensional (2D) (or three dimensional [3D]) images are reconstructed along one or two axes parallel to the acquisition surface (named cross-range directions) as well as the direction along the unit normal of the surface (the range direction). On the other hand, true holography utilizes measurements over a 2D plane and the image coordinates are always along the cross-range directions.

2.1 Microwave Hologram Formation

In general, hologram formation means detecting the phase and amplitude of the field scattered by the object and storing the data. In optical holography, the hologram is essentially a planar record of the intensity of an interference pattern obtained in a single-frequency measurement with two coherent waves. In an optical hologram, the record is imprinted on a photographic plate (the recording medium). In microwave holography, although both phase and amplitude can be measured, in early works, mainly the intensity pattern was acquired. This is referred to as "indirect microwave holography" since the phase information is extracted indirectly from the intensity measurements. Later, in particular in the quasi-microwave holography discussed in the next chapters, both the phase and amplitude of the scattered waves are measured. This is referred to as "direct microwave holography" since the phase of the scattered wave is measured directly. What is important to realize is that in both direct and indirect microwave holography, both magnitude and phase information can be obtained from the hologram and they are exploited in order to generate images of the objects and to achieve resolution enhancement compared to the 2D images obtained from amplitude measurements only. In fact, in imaging instrumentation, holography is defined as an interferometric technique for recording the amplitude and the phase of a wave.

In the following, we briefly review the expressions for the intensity pattern over the hologram plane obtained in indirect microwave holography. Intensity hologram measurements require a reference wave coherent with the scattered wave due to the object. The coherent reference is required to obtain the object wave's phase, which is necessary for reconstructing the object and forming the

Figure 2.1 Illustration of hologram recording step in true microwave holography. A transmitter antenna illuminates the object and another reference antenna radiates a reference wave coherent with the illuminating wave. Both radiated waves can be produced by the same microwave source connected to a directional coupler, a form of beam splitter. The intensity of the wave can be then measured on the hologram plane [85].

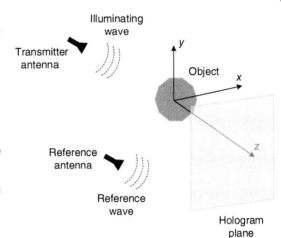

images. To describe the measurements, we utilize scalar notation because in many cases the field is linearly polarized, and holograms are usually flat.

The holographic recording step is illustrated in Figure 2.1. Two coherent waves, the illuminating waves and the reference waves, are obtained from a microwave monochromatic generator by splitting the generator's output with a directional coupler or a power splitter. The illuminating field E^{inc} is scattered due to the presence of the object producing E^{sc}. This scattered field interferes with the reference field E^{ref}. The reference field must not interact with the object.

The interference pattern is recorded on a plane (hologram recording step). In general, the hologram plane can be positioned at any desired angle with respect to the illumination as long as it captures the sufficient and useful portion of E^{sc} without being exposed directly to the illuminating wave.

The recording medium or probe on the hologram plane records the intensity distribution:

$$I_H = \left(E^{\text{sc}} + E^{\text{ref}}\right)\left(E^{\text{sc}} + E^{\text{ref}}\right)^* = \left|E^{\text{sc}}\right|^2 + \left|E^{\text{ref}}\right|^2 + \left(E^{\text{sc}}\right)^* E^{\text{ref}} + E^{\text{sc}}\left(E^{\text{ref}}\right)^*$$

$$(2.1)$$

where $(.)^*$ denotes the conjugation operator. It is clear from (2.1) that the hologram contains the amplitude and the phase information of E^{sc} in the third and the fourth terms. If the reference field E^{ref} is known, then E^{sc} can be extracted from each one of these terms. This is the task of the reconstruction step in holography which will be described shortly.

Besides, to explicitly observe the phase terms in the measured intensity distribution I_H, we can write the scattered and reference fields as:

$$E^{\text{sc}} = a_{\text{sc}} \exp(i\phi_{\text{sc}})$$

$$(2.2)$$

$$E^{\mathrm{ref}} = a_{\mathrm{ref}} \exp(i\phi_{\mathrm{ref}}) \tag{2.3}$$

where a_{sc} and a_{ref} are the amplitudes of the scattered and the reference waves, respectively, and ϕ_{sc} and ϕ_{ref} are phases of the scattered and reference waves, respectively. From (2.1)–(2.3), the intensity distribution can be written as:

$$I_H = a_{\mathrm{sc}}^2 + a_{\mathrm{ref}}^2 + a_{\mathrm{sc}} a_{\mathrm{ref}} \exp[i(\phi_{\mathrm{sc}} - \phi_{\mathrm{ref}})] + a_{\mathrm{sc}} a_{\mathrm{ref}} \exp[-i(\phi_{\mathrm{sc}} - \phi_{\mathrm{ref}})].$$

$$\tag{2.4}$$

From (2.4) it is observed that the phase of the scattered wave due to the object ϕ_{sc} is implicit in the observable values of I_H. This causes a pattern of intensity fringes. The effect of phase variations is to modulate fringe positions. These relationships were first presented by Gabor.

2.2 Microwave Detectors and Sampling Methods for Intensity Hologram Measurements

Detectors that have been used for intensity measurements at microwave frequencies can be categorized in two classes: (i) antennas that can be employed along with the receiver circuits and (ii) continuous materials that change under microwave illumination.

Liquid crystal films and Polaroid films are examples of continuous detectors. Liquid crystals can be utilized because their color changes with temperature (e.g. see [111]). If liquid crystal is applied to a layer that absorbs the incident microwave energy it gets heated and its color changes. Subsequent photographic reduction of the displayed color patterns through colored filters produces black and white holograms suitable for reconstruction with laser light. In [112], real-time imaging of concealed objects has been suggested without an image reconstruction phase. A liquid crystal film has been utilized to record the scattered field near an object and obtain its geometric projection. Although liquid crystal detectors are feasible, their sensitivity is low. In [113], polaroid films have been utilized by illuminating the film with microwaves after exposing it to white light to start the development process. Microwaves heat the film and change the development rate so that variations in microwave intensity produce distinct colors. There, again sensitivity is low.

In a common sampling scheme, a receiving antenna scans while the transmitting antenna, the reference antenna, and the object remain stationary. A diode detector or receiver measures the intensity distribution formed by the interfering scattered and reference waves. Although economical, a radiated reference wave has the disadvantage that it can be scattered by the surroundings or the probe support leading to distortions in the reference field. To avoid this problem, the reference wave can be directed to the detector through waveguide.

This choice usually requires additional equipment, rotating joints, or flexible cables to convey the reference signal to the detector or mixer. It also requires a phase shifter to change the phase linearly with the probe motion to simulate an inclined reference beam.

The intensity scattered by an object can be measured by an array of antennas with crystal diode detectors. The drawback is that an array is expensive and dense sampling of the intensity is not possible due to the large size of the antennas. Thus, data often are measured by mechanically scanning an area with one antenna. Scanning causes delays which are undesirable since instrument or source instabilities cause errors. Besides, delays are unacceptable for recording transient events. Nevertheless, scanning is adequate for laboratory measurements or for nondestructive testing and similar situations where the subject being imaged does not change with time.

Single antennas have been scanned in several ways. Linear motion on a series of parallel lines is common (e.g. see [97]). Parallel arcs have been used by placing the antenna on an arm with fixed radius from a center that is laterally displaced. Also, a spiral path can be utilized by rotating an arm about a fixed center and decreasing the radius during the rotation. Double circular scans have also been considered to accelerate the mapping of the microwave fields over an aperture.

Measurements can be accelerated by scanning a linear array of antennas and crystal detectors over the recording aperture. The array is translated in the direction perpendicular to its length to map the hologram data in a rectangular aperture. An approach that compromises between cost and speed utilizes two crossed linear arrays.

Measurements of both amplitude and phase require a receiver that is more complex than an intensity detector. With modern microwave instruments, e.g. vector network analyzers (VNAs), we can easily obtain both the amplitude and the phase of E^{sc} (direct microwave holography). Thus, the recording step is simpler to implement. This will be described in more detail when we describe the direct microwave holography in the next chapters. The advantages of using intensity-only measurements (indirect microwave holography), however, is that intensity microwave detectors are cheaper and often smaller in size. It is worth noting here that when only intensity is measured, sampling theory shows that the distance between adjacent samples must be half that for phase and intensity measurements. The intuitive reason is that a measurement of phase and intensity gives two real numbers; intensity measurement gives one. Criteria for sampling have been discussed by Lohmann [114].

The use of an oscilloscope has also been proposed to record the intensity distribution when using a single scanning antenna. It has been considered as an inexpensive approach to measure the fringe patterns (e.g. see [97]). In this approach, the detected voltage by the oscilloscope is proportional to microwave intensity. This is because the microwave signal modulates the electron beam and the spot brightness. The beam is moved synchronously with the probe by a servomechanism. In this setup, the phase can be measured and in such case, fringes can be formed by modulating the brightness with a voltage proportional to the phase. If phase and amplitude are measured, both quantities can be displayed by forming detour phase holograms with an oscilloscope. In all these cases, the oscilloscope displays are photographed and the size of the photographs is reduced to form transparencies that are holograms. Images are produced by illuminating the holograms with laser light.

Although an oscilloscope display is convenient for a single probe, it is not suitable to be used with a linear array of receiving antennas since an array would require a multiple beam oscilloscope or multiplexing the array output and displacing the oscilloscope beam. It is simpler to utilize a linear array of light sources, using a source for each antenna and detector. The voltages from the detector are amplified to control the brightness of the source. A time exposure photograph of the array of light sources can be made while a linear array is scanned in the direction perpendicular to its length. However, dynamic range would be limited in this approach. In an early work, fringes were displayed with a lightbulb and the bulb was photographed as it scanned the field [115].

Additionally, microwave holograms can be formed by measuring the fringe patterns and using them as guides for fabricating holograms consisting of metallic scatterers or absorbing material. Phase data can also be utilized to form phase holograms by contouring dielectrics. Microwave hologram data can then be digitized for computer processing. For the lower microwave frequencies, digital processing is superior to experimental reconstruction with light because it eliminates the scaling problem. For shorter millimeter waves the production of an optical hologram requires less scaling.

2.3 Wave Front Reconstruction

Because a hologram contains information derived from the Fraunhofer or Fresnel wavefields diffracted by the object, wave front reconstruction and image retrieval can be described by Fourier transform (FT) operations. These operations can be performed either experimentally, say with an analog optical computer that may utilize the FT property of a convergent lens, or digitally by a computer, usually employing fast Fourier transform (FFT) algorithms. Both methods have advantages and disadvantages that have made one better suited than the other depending on the application and also considering the technology limitations at the time of their development.

The advantages of optical computing include relative simplicity and instantaneous parallel processing of 2D input data at the speed of light. The limitations of optical computing, in particular at the time of their initial development, stemmed from the slow rates at which data could be put into the computer, usually by means of photographic film. Digital computing had the advantage of flexible processing, but it could only handle the data serially. Although decades ago the cost of nearly real-time digital computation of hologram data was high, nowadays, the availability of powerful computers is cutting such costs significantly.

In the reconstruction step of optical holography, the processed photographic hologram is illuminated by the coherent reconstruction beam, which is usually (but not necessarily) of the same frequency as the reference field E^{ref}. In this step, the object is absent and the illuminating beam is off but the same geometrical arrangement of the hologram and the reference beam is used as that at the time of the hologram recording.

Upon passing through the hologram, the reconstruction beam acquires the phase and amplitude modulation of the original wave, the intensity pattern of which created the hologram. The wave emerging from the illuminated hologram is mathematically described by

$$E_I = E^{\mathrm{ref}} I_H = E^{\mathrm{ref}} \left| E^{\mathrm{ref}} \right|^2 + E^{\mathrm{ref}} \left| E^{\mathrm{sc}} \right|^2 + \left(E^{\mathrm{sc}} \right)^* \left(E^{\mathrm{ref}} \right)^2 + E^{\mathrm{sc}} \left| E^{\mathrm{ref}} \right|^2. \qquad (2.5)$$

The first term is the uniform reference wave, which does not carry any information about the object. The second term can be made negligibly small compared to the others by making the reference field much stronger than the scattered field. The last term is proportional to the original scattered field produced by the object during the recording step. In optical holography, with the off-axis reference beam proposed by Leith and Upatnieks [82, 93], this wave can be focused in a direction different from that corresponding to the third term, which is proportional to the conjugated scattered field. Note that in the original arrangement of Gabor, there was no reference beam, or, equivalently, one could think of a system where the illuminating and reference beams were coincident.

In the original microwave holography, the reconstruction step closely resembled that in optics. The microwave hologram was scaled down to a size that fits on a photographic hologram and was subsequently processed using laser light (e.g. see [99]).

Another concept for a real-time microwave holographic camera has been proposed in [85] and is shown in Figure 2.2. The camera combines a 2D hologram recording array of sensors with real-time optical reconstruction scheme. The sensor array converts the intensity distribution in the hologram fringe pattern into electronic, intermediate frequency signals that are amplified and processed in parallel, and used to activate an array of light emitting diodes. Using this configuration allows for instantaneous visualization of the microwave hologram fringe pattern. Glow discharge detectors were suggested for experimental hologram recording arrays because of their extremely low cost, their wide spectral response (covering the microwave and millimeter wave range), and because they function as an antenna and transducer

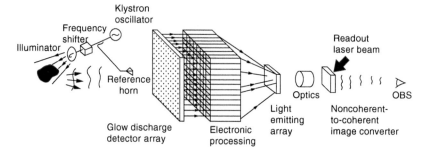

Figure 2.2 Concept of real-time microwave holographic camera proposed in [85].

simultaneously. Nowadays, the tasks performed by the optics are emulated by computer algorithms described in the following section.

2.4 Recent Indirect Holographic Imaging Techniques

As described before, the so-called indirect microwave holography is a holographic technique in which the phase of the field is extracted via the measurement of scalar values (field intensity). On the other hand, in direct microwave holography, the phase of the field is measured directly using instruments capable of vector (i.e. concurrent magnitude and phase) measurements.

In addition to imaging, indirect microwave holography has been utilized for determining the antenna radiation patterns via the reconstruction of the complex near-zone and aperture fields of antennas (e.g. see [116]). Since vector measurement of the antennas is complicated and expensive, this approach provides an alternative measurement technique to reconstruct the complex field values from scalar measurements in a simple and inexpensive manner. In this technique, the antenna under test acts as a transmitter while a single probe antenna is used to record the near-field over a plane and provide a hologram. The measured quantity can be the received power by the probe antenna. The reference wave is obtained by splitting and combining a portion of the signal fed to the antenna under test with the near-field signal of that antenna by using a transmission line, a phase shifter, and possibly a variable attenuator. This technique is then capable of reconstructing the complex fields from the measured scalar intensity patterns, without the need for nonlinear inverse-scattering methods.

In [117], the indirect holographic technique for antenna pattern measurements has been extended to the imaging of objects. There, scalar holographic patterns have been recorded and processed to recover the complex scattered field (due to an object) over the measurement plane (hologram). Then, back-propagation has been utilized to obtain images of the objects. Such images show the variations of the object's reflectivity, or, equivalently, the distribution of its scattering centers. Thus, they are qualitative rather than quantitative. Below, we describe this technique in more detail.

2.4.1 Producing Reference Signal with a Linear Phase Shift

Figure 2.3 illustrates the experimental setup for indirect microwave holographic imaging [117]. In this setup, the object to be imaged is illuminated by a small antenna, Tx. The scattered field due to the object $E^{sc}(x, y)$ is received by the probe antenna Rx, at the point x,y on the measurement plane, $z = d$, and is measured by the power meter. The measurement plane is located in the radiating near-field and sufficiently far from the object (such as $z = 3\lambda$ to 5λ) to ensure that the evanescent fields vanish. In order to accurately recover the scattered field, the probe antenna needs to provide coverage across the scanning aperture without having nulls in the radiation pattern that prevent receiving the waves. Open-ended waveguide probes are good candidates [117].

The received signal by the probe antenna is applied to one of the inputs of the hybrid tee. A phase-coherent reference signal, $E^{ref}(x, y)$, is applied to the second input of the hybrid tee. The input to the power meter is taken from the sum port of the hybrid tee with the difference port terminated with a matched load. The reference signal is assumed to have a uniform amplitude and a linearly increasing phase shift, k_r, which can be applied along the x direction, $k_r = k_x$, the y direction, $k_r = k_y$, or a combination of both, $k_r = \sqrt{k_x^2 + k_y^2}$.

To simplify the explanation of the technique, without loss of generality, let us consider a 1D scan along the x direction. In this case, the reference wave is of the form

$$E^{ref}(x) = E_0 e^{-jk_r x} \tag{2.6}$$

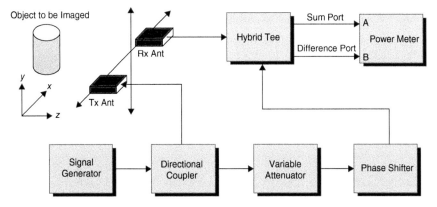

Figure 2.3 Indirect holographic measurement system [117].

where

$$k_r = \frac{\Delta\phi}{\Delta x} \tag{2.7}$$

where k_r is the phase gradient and $\Delta\phi$ is the phase shift applied for each sampling step Δx, i.e. from one sampling position to the next.

If the scattered field due to the object along the x axis is denoted by $E^{sc}(x)$, the intensity of the output signal from the sum port of the hybrid tee will be

$$I_H(x) = \left|E^{sc}(x) + E^{ref}(x)\right|^2 = \left|E^{sc}(x)\right|^2 + \left|E^{ref}(x)\right|^2$$
$$+ \left(E^{sc}(x)\right)^* E^{ref}(x) + E^{sc}(x)\left(E^{ref}(x)\right)^*. \tag{2.8}$$

The scalar output of the power meter is proportional to $I_H(x)$. In other words, $I_H(x)$ represents the values of the hologram along the x axis. In order to obtain the complex values of the scattered field over the hologram, in the next step, we take the FT of both sides of (2.8) to obtain

$$\mathrm{FT}_{1D}\{\tilde{I}_H(x)\} = \mathrm{FT}_{1D}\{|E^{sc}(x)|^2\} + \mathrm{FT}_{1D}\{|E^{ref}(x)|^2\}$$
$$+ \mathrm{FT}_{1D}\{(E^{sc}(x))^*\}\otimes\mathrm{FT}_{1D}\{E_0 e^{-jk_r x}\} + \mathrm{FT}_{1D}\{E^{sc}(x)\}\otimes\mathrm{FT}_{1D}\{E_0 e^{+jk_r x}\} \tag{2.9}$$

where $\mathrm{FT}_{1D}\{\cdot\}$ denotes the 1D FT operator and $\otimes$ denotes the convolution operator in the Fourier domain. The first two terms of this expression are centered at the origin in the spatial Fourier domain (k_x) whereas the third and fourth terms are displaced from the origin by $\pm k_r$. This is illustrated in Figure 2.4.

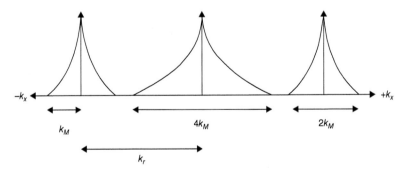

Figure 2.4 The spatial-frequency spectrum of the intensity pattern recorded at the sum port of the hybrid tee [117].

Here, we assume that the scattered field has a spectrum limited within the spatial-frequency range $-k_M < k_x < k_M$, i.e. k_M is the spectral extent of the scattered field. We also assume that the phase gradient applied to the offset reference signal is such that $k_r > 3k_M$. In such cases, the third and fourth terms of (2.8) can be separated and the other terms can be removed by filtering. To filter the unwanted terms, various windows can be employed in the Fourier domain. The best window is usually selected empirically for the particular setup [117]. The simplest window is a rectangular function in the spatial Fourier domain. The properties of the filter were considered in [118] and are summarized here. First, the filter should have a sharp cutoff wave number, above which the filter stopband attenuation becomes very high, e.g. greater than 40 dB. Second, the variation the of filter in the spatial domain should be in the form of a 2D function which is separable, i.e. it should be of the form $h(x,y) = h_1(x)h_2(y)$. It should be noted that the separability of the filter function provides ease in computation and that it is not a theoretical requirement. Third, in the spatial domain, the filter function should be spatially limited, i.e. $h(x, y) \cong 0$ for $x \geq x_m$ or $y \geq y_m$. This requirement is necessary for the extracted scattered field to be a spatially limited function. In [118] Blackman filter has been analyzed since it satisfies the above-mentioned three requirements. In [116], Hamming-window has been employed. This is more suitable for near-field reconstruction. Even though the Blackman filter appears to have better properties, the Hamming filter is simpler from a numerical point of view.

Let us assume that we aim at extracting the fourth term via filtering and also let us assume that we are able to extract this term perfectly. This produces a filtered signal of the form

$$\text{FT}_{1D}\left\{\hat{I}_H(x)\right\} = \text{FT}_{1D}\left\{E^{sc}(x)\right\} \otimes \text{FT}_{1D}\left\{E_0 e^{+jk_r x}\right\}. \tag{2.10}$$

The resultant signal consists of the FT of the original scattered field shifted in the spatial-frequency domain by k_r. This spatial-frequency offset can be readily removed and the inverse Fourier transform can be applied to obtain

$$\hat{E}^{sc}(x) = \text{FT}_{1D}^{-1}[\text{FT}_{1D}\{E^{sc}(x)\}] \tag{2.11}$$

where $\hat{E}^{sc}(x)$ is the recovered (estimated) complex-valued scattered field at the measurement plane (hologram).

In order to reduce the overlap between the terms in the Fourier domain, both the sum and difference signals that are available from the two output ports of the hybrid tee can be employed [117]. At the sum port, the measured intensity is obtained from the expression given in (2.8). The intensity of the signal available from the difference port will be of the form

$$\begin{aligned} I^-(x) &= \left|E^{sc}(x) - E^{ref}(x)\right|^2 = \left|E^{sc}(x)\right|^2 + \left|E^{ref}(x)\right|^2 \\ &\quad - \left(E^{sc}(x)\right)^* E^{ref}(x) - E^{sc}(x)\left(E^{ref}(x)\right)^*. \end{aligned} \tag{2.12}$$

Subtracting (2.12) from (2.8) produces the following signal:

$$I^+(x) - I^-(x) = 2\left[(E^{sc}(x))^* E^{ref}(x) - E^{sc}(x)(E^{ref}(x))^*\right]. \tag{2.13}$$

In (2.13) the terms that are centered at the origin in the spatial-frequency domain have been removed. This significantly increases the available spatial-frequency bandwidth before overlapping of terms can occur and also allows increased sample spacing since k_r in (2.7) can be large.

The extracted complex-valued scattered field at the measurement plane can be back-propagated to the position of the objects as:

$$\hat{E}^{sc}(x,z=0) = \frac{1}{2\pi}\int FT_{1D}\{\hat{E}^{sc}(x)\}e^{jk_z d}e^{-jk_x x}dk_x. \tag{2.14}$$

$\hat{E}^{sc}(x,z=0)$ is the estimate of the scattered wave at the position of the object and it can be plotted as a qualitative image of the object. As mentioned earlier, similar method has been implemented for the determination of antenna aperture fields. It is worth noting that this back-propagation method is limited to the case where the medium is linear and homogeneous.

Although the off-axis indirect holography, where the reference signal is introduced with a linear phase shift, has found widespread use at optical frequencies, its use at microwave frequencies has been limited. This is due to the practical difficulties in producing the required reference signal with a linear phase gradient.

2.4.2 Sample Imaging Results

In [117], an imaging experiment has been conducted on a thin rectangular aluminum sheet. In this experiment, the size of the sheet is 120 mm × 200 mm and it is located 120 mm away from the scanning aperture, as shown in Figure 2.5. The frequency of operation is 12.5 GHz. Measurements are taken at sample spacing of $\Delta x = \Delta y = 6$ mm, which corresponds to $\lambda/4$ over a rectangular scanning aperture of 450 mm × 450 mm. The offset reference signal is synthesized by introducing a phase shift of $\Delta\phi = 2\pi/3$ rad between the sample spacing Δx. This produces an offset reference wave vector of $k_r = 8\pi/3\lambda$ rad/m ($k_0 = 2\pi/\lambda$ rad/m). The sharp edges of this object are of particular interest since the quality of the reconstructed image at those edges shows the effects of term overlap in the spatial Fourier domain.

The holographic intensity pattern available from only the sum port of the hybrid tee has been selected. This is to investigate the effects of the possible overlap of terms in the spatial Fourier domain. Figure 2.6 shows the recorded holographic intensity pattern over the 450 mm × 450 mm scanning aperture, with a linear phase shift applied along the horizontal axis. The horizontal and vertical axes are shown in terms of the sample number. The periodic

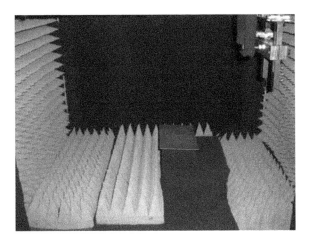

Figure 2.5 Holographic imaging of a rectangular metal plate [117].

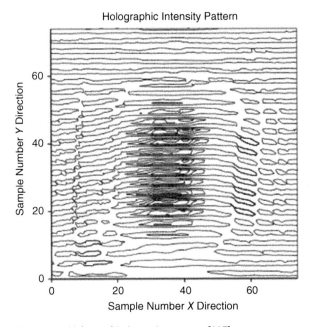

Figure 2.6 Holographic intensity pattern [117].

variations in this pattern occurring every third sample are a direct result of applying the phase shift of $\Delta\phi = 2\pi/3$ rad between samples.

With the sample spacing of $\Delta x = \Delta y = 6$ mm along x and y, a 2D array of size 75×75 samples is recorded containing the intensity values. This is zero-filled to provide a 256×256 data array before taking FT. Taking FT of the recorded intensity provides the pattern shown in Figure 2.7, where the axes are given in terms of the spatial-frequency sample number, with $\Delta k_x = \Delta k_y = 4.09$ rad/m.

A linear phase gradient of

$$k_r = \frac{\Delta \phi}{\Delta x} = \frac{4k_0}{3} \tag{2.15}$$

was used. The phase gradient introduced into the reference signal provides an offset wave vector, k_r, that would ensure separation of terms in the spatial Fourier domain, provided that the spatial-frequency spectrum of the scattered wave k_M satisfies $k_M < k_r/3$. In general, this could be achieved by applying the linear phase shift along the x axis, the y axis, or as a combination of both.

The FT shown in Figure 2.7 was obtained using only the intensity pattern available from the sum output of the Hybrid Tee and as given by (2.8). Figure 2.8 shows the plot produced by a cut through the central section of the 2D FT pattern in Figure 2.7. It demonstrates that for this object, good separation could be obtained between terms when using only the sum intensity pattern. Also, Figure 2.9 shows the plane-wave spectrum (PWS) of the field scattered from the metal plate.

Following the procedure outlined before, the complex scattered field at the measurement plane can be reconstructed. The magnitude and phase of the reconstructed scattered field from the metal plate at the measurement plane, $z = 0$, are shown in Figure 2.10. There, the horizontal and vertical axes represent the original physical aperture of 450 mm × 450 mm. The outline of the original aluminum plate has been shown by dashed line in the same figure. Figure 2.10 shows that the reconstructed magnitude image at the measurement plane

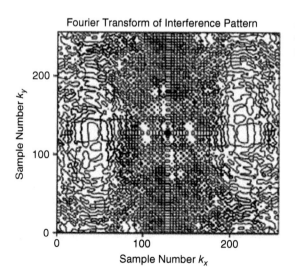

Figure 2.7 Fourier transform of the holographic intensity pattern [117].

Figure 2.8 A cut through of the Fourier transform to demonstrate the separation of terms [117].

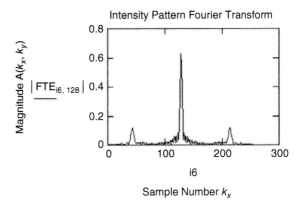

Figure 2.9 The plane-wave spectrum of the metal plate: (a) along k_x and (b) along k_y [117].

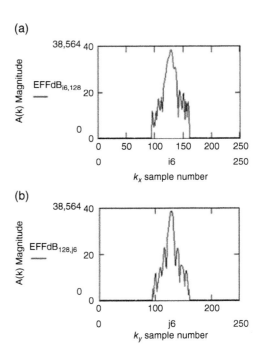

provides a defocused image of the object that lacks some of the edge details of the original object. The phase image gives an outline closer to the original rectangular metal plate.

Additionally, the reconstructed complex field at the acquisition aperture can be back-propagated to the plane of the object, using (2.14), to produce an image of the original scattering object, as shown in Figure 2.11. The outline of the

(a)   (b)

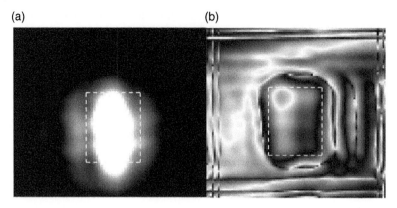

Figure 2.10 Reconstructed complex scattered field at the measurement plane:
(a) reconstructed magnitude and (b) reconstructed phase [117].

(a)   (b)

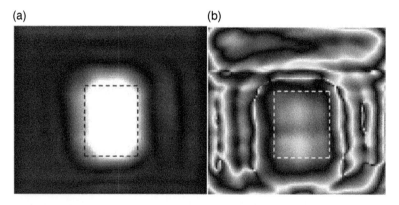

Figure 2.11 Reconstructed complex scattered field at the object plane: (a) reconstructed
magnitude and (b) reconstructed phase [117].

original aluminum plate is shown by dashed line in the same figure. The exam-
ination of these images shows good agreement with the outline of the actual
aluminum plate. As before, the horizontal and vertical axes of these figures represent the original measurement aperture of 450 mm × 450 mm.

3

Direct and Quasi-Microwave/Millimeter-Wave Holography for Far-Field Imaging Applications

The techniques discussed in this chapter are known as direct holographic imaging methods since the phase of the scattered field is measured directly via vector measurement. Besides, the three-dimensional (3D) holographic imaging discussed in this chapter can be referred to as quasi-microwave holography. In the original quasi-microwave holography, pulsed or multi-frequency radiation is utilized to illuminate the object and the measurements are performed along a line. The two-dimensional (2D) images are then reconstructed along one axis orthogonal to the radial direction (cross-range direction) and one axis along the radial direction (range direction). In later developments of quasi-microwave/millimeter-wave holography, wideband scattered field is sampled over a rectangular or cylindrical surface and 3D images are reconstructed. In this chapter, we describe how these 3D quasi-microwave holography techniques emerged as a combination of 2D synthetic aperture radar (SAR) imaging and 2D microwave holographic imaging. Pacific Northwest National Laboratory (PNNL), Richland, WA, pioneered the development of these techniques. There, millimeter waves were employed for detection of weapons or dangerous items hidden beneath clothing. This approach gained significant attention and led to the development of concealed weapon detection systems that are currently used in numerous airports and public enterprises. Below, we first briefly describe the history of using microwave and millimeter-wave holography for concealed weapon detection and then we continue with the relevant mathematical developments.

3.1 Using Microwave and Millimeter-Wave Holography for Concealed Weapon Detection

In general, the applications related to security screening have been the major motivation for developing microwave and millimeter-wave holographic imaging. Millimeter-wave holographic imaging for concealed weapon detection

Real-Time Three-Dimensional Imaging of Dielectric Bodies Using Microwave/Millimeter-Wave Holography, First Edition. Reza K. Amineh, Natalia K. Nikolova, and Maryam Ravan.
© 2019 by The Institute of Electrical and Electronics Engineers, Inc.
Published 2019 by John Wiley & Sons, Inc.

was originally proposed in [84]. There, the imaging technique utilized a stationary source and a scanned receiver system that employed optical (film-based) reconstruction. This technique was significantly improved in [119] by utilizing a scanned transmitter and receiver configuration and digital reconstruction. This configuration produced high-quality imagery because the object was illuminated over a broad range of angles, both vertically and horizontally. In the 2D microwave holographic imaging in [119], an array of sequentially switched transmitter and receivers are scanned quickly over a large aperture to actively illuminate the object. This system operates at a single frequency and is coherent, which means that both the phase and amplitude of the reflected signal are recorded. Thus, this approach can be considered as a direct microwave holography. It can be also considered as a true microwave holography since the scattered waves are measured over the 2D aperture to produce an image over another plane parallel to it and passing the object. The coherent data can be mathematically reconstructed in a computer to form a focused image of the object without the need for a lens. Advantages of this technique include: (i) near real-time operation; (ii) high-resolution; (iii) computer reconstruction allows focusing at any depth (imaging distance from the object to the scanned aperture); and (iv) large aperture (in concealed weapon detection, full-body can be fitted within the field of view). Despite the above-mentioned advantages, the disadvantage of this imaging system is that the range of focusing depth is short. This is due to the close-range operation which makes the system sensitive to small variations in the imaging distance. On the other hand, the imaging aperture and the field of view (image size) is large and may contain variations of the object surface, i.e. the imaging distance (depth of focus) over the field of view may be variable. Therefore, the image of an object with significant depth variations, such as the human body, cannot be reconstructed well with a single depth of focus. Thus, in [5], the holographic imaging system was extended from single-frequency operation to wideband (many frequencies) operation. Rather than simply allowing the formation of a 2D image reconstructed at each operation frequency, wideband imaging allows the formation of a fully focused 3D image from wideband data gathered over a 2D aperture (quasi–millimeter-wave holography). This advancement overcomes the focusing limitation present in the narrow-band systems. As explained in the rest of this chapter, this algorithm extends the work reported in [120–122], which discusses the algorithms for 2D SAR imaging.

3.2 Monostatic 2D SAR Imaging

First, we describe the 2D SAR imaging technique presented in [120]. In the theoretical discussions presented in this section, (x, y) and (k_x, k_y) denote the spatial and the spatial-frequency variables, respectively. Here, the x coordinate is used to identify the range position and y specifies the cross-range position. For a spatial domain function, e.g. $f(x, y)$, its spatial Fourier transform (FT) is denoted by $\tilde{F}(k_x, k_y)$. The imaged object is assumed to be within a disk of radius X_0, centered at the origin in the spatial domain. The surrounding medium is homogeneous and the speed of propagation in the medium is c. The wave number of a wave with temporal frequency ω traveling in this homogeneous medium is denoted by $k = \omega/c$.

We assume that the object is illuminated with a transmitter antenna operating at multiple frequencies. We also assume that the complex-valued reflected signal is measured with the same antenna, i.e. both phase and amplitude of the scattered field are measured using proper data acquisition circuit.

Suppose that the radar moves along the line $x = X_1$ on the (x, y) plane. The radar makes a transmission and its corresponding reception at $(X_1, Y_1 + u)$ where $u \in [-L, L]$ and $2L$ is the size of the synthesized linear aperture. (X_1, Y_1) are known constants. Also, the spatial-frequency variable corresponding to u is denoted by k_u.

Here, we assume that the antenna is an isotropic source/receiver, i.e. it transmits and receives waves uniformly in all directions. Thus, the round-trip phase delay of the echoed signal by a point scatterer at (x, y) is $2k\sqrt{(X_1 - x)^2 + (Y_1 + u - y)^2}$ and the total complex-valued echoed signal at the receiver becomes

$$s(u, \omega) = \int \int f(x, y) \cdot \exp\left[j2k\sqrt{(X_1 - x)^2 + (Y_1 + u - y)^2}\right] dx dy \tag{3.1}$$

where $f(x, y)$ is the object's reflectivity function. The free-space loss factor $1/[(X_1 - x)^2 + (Y_1 + u - y)^2]$ for amplitude decay on the right side is suppressed. This is due to the assumption that in SAR imaging the range is significantly greater than the object's size. This makes the variations of the amplitude of the reflected wave from various points of the object negligible. Moreover, the spherical wave that appears on the right side of (3.1) can be decomposed in terms of plane waves as follows:

$$\exp\left[j2k\sqrt{(X_1 - x)^2 + (Y_1 + u - y)^2}\right]$$
$$= \int \exp\left[j\sqrt{4k^2 - k_u^2} \cdot (X_1 - x) + jk_u(Y_1 + u - y)\right] dk_u \tag{3.2}$$

See appendix A in [120] for a discussion on (3.2). Note that the integral in (3.2) is in the complex spatial-frequency domain, wherein the spatial-frequency domain amplitude function $1/\sqrt{4k^2 - k_u^2}$ has been suppressed.

It is worth noting that since this imaging system was proposed for SAR imaging (far-field imaging), here, we only consider the real values of $k_u \in [-2k, 2k]$. This indicates that only the propagating (non-evanescent) components of the reflected waves are taken into account (evanescent components that correspond to values of k_u, where $|k_u| > 2k$, decay very fast and can be only measured in close proximity to the object).

Using (3.2) in (3.1), and after some rearrangement, one obtains

$$s(u,\omega) = \int \exp\left[j\left(\sqrt{4k^2 - k_u^2} \cdot X_1 + k_u Y_1\right)\right] \cdot$$
$$\left\{\int\int f(x,y) \exp\left[-j\left(\sqrt{4k^2 - k_u^2} \cdot x + k_u y\right)\right] dx dy\right\} \cdot \exp\left(jk_u u\right) dk_u \tag{3.3}$$

where the term inside the brackets with the double integral can be written as a 2D FT with spatial-frequency variables $\sqrt{4k^2 - k_u^2}$ and k_u. It is denoted as $\tilde{F}\left(\sqrt{4k^2 - k_u^2}, k_u\right)$. Thus, (3.3) can be written as

$$s(u,\omega) = \int \exp\left[j\left(\sqrt{4k^2 - k_u^2} \cdot X_1 + k_u Y_1\right)\right] \cdot \tilde{F}\left(\sqrt{4k^2 - k_u^2}, k_u\right) \cdot \exp\left(jk_u u\right) dk_u. \tag{3.4}$$

We denote the spatial FT of $s(u, \omega)$ with respect to u by $\tilde{S}(k_u, \omega)$. Taking the spatial FT of both sides of (3.4) with respect to u yields the following inverse equation:

$$\tilde{F}\left(\sqrt{4k^2 - k_u^2}, k_u\right) = \exp\left[-j\left(\sqrt{4k^2 - k_u^2} \cdot X_1 + k_u Y_1\right)\right] \cdot \tilde{S}(k_u, \omega) \tag{3.5}$$

or

$$\tilde{F}(k_x, k_y) = \exp\left[-j(k_x X_1 + k_y Y_1)\right] \cdot \tilde{S}(k_u, \omega) \tag{3.6}$$

where k_x and k_y are defined as

$$k_x = \sqrt{4k^2 - k_u^2} \tag{3.7}$$

$$k_y = k_u. \tag{3.8}$$

The inversion in (3.6) indicates that the samples of the 2D FT of the object's reflectivity function, i.e. $\tilde{F}(k_x, k_y)$, can be obtained from the one-dimensional (1D) FT of the complex-valued echoed signal, i.e. $\tilde{S}(k_u, \omega)$. This procedure is also referred to as Doppler processing across the synthetic aperture which indicates data transformation from u to k_u.

3.3 Development of 3D Quasi-Holographic Imaging as a Combination of Monostatic 2D SAR Imaging and True 2D Holographic Imaging

The image reconstruction algorithm derived in this section is a combination of the true holographic image reconstruction (backward-wave propagation) with a monostatic SAR image reconstruction algorithm discussed in Section 3.2. As discussed in Section 3.2, a wideband SAR imaging algorithm can be employed to reconstruct a 2D image from data collected over a linear aperture. The derivation presented in this section extends that work by making the aperture planar instead of linear, which allows for a full 3D image reconstruction. Similar to the work discussed in Section 3.2, slowly varying amplitude functions are ignored in the derivation, as they will not have a significant effect on the reconstructed images. Besides, the wideband reconstruction algorithm can be considered to be an extension of the single-frequency "backward-wave" 2D holographic image reconstruction algorithm (true holographic imaging) to 3D imaging by adding the range dimension. In the following, we first describe the holographic 2D imaging and then proceed with extending that to 3D imaging.

3.3.1 Single-Frequency Holographic 2D Imaging

True holographic 2D imaging is a means of forming focused images of the objects from coherent-wave data gathered remotely over a 2D aperture. A common arrangement uses a planar aperture where the same antenna illuminates the object and then receives the reflected waves, which are recorded coherently, digitized, and stored in the computer. The system could be also quasi-monostatic, which means that the transmitting and receiving antennas are separate, but in approximately the same location and may be assumed to be coincident at the midpoint between the two antennas. The data are mathematically processed on the computer to form a focused image of the object's reflectivity function. Similar to the SAR imaging described in Section 3.2, slowly varying amplitude functions are typically ignored in the derivation, as they will not have a significant effect on the reconstructed images. Besides, the technique discussed here relies extensively on the use of FTs that may be computed very efficiently using the fast-Fourier-transform (FFT) algorithm. The limitations on the resolution in this 2D imaging technique stem from the diffraction limit imposed by the wavelength. The resolution is also affected by the antenna beamwidth, the size of aperture, and the distance to the object. The reconstructed image is in focus only over a shallow depth which is determined by the aperture size or the antenna beamwidth [5].

The measurement configuration is shown in Figure 3.1. The source is assumed to be at position (x', y', z_0), and a general point in the object is assumed to be at position $(x, y, 0)$. The object is assumed to be characterized by a

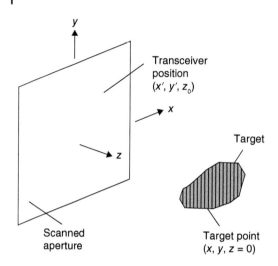

Figure 3.1 True holographic 2D imaging system [5].

reflectivity function $f(x, y, z)$, which is simply the ratio of the reflected field to the incident field.

The response measured at the transceiver is represented as the superposition of the backscattered waves generated by each point on the object. These are modeled by a simple multiplication of the object's reflectivity and the round-trip phase delay associated with the signal path from the transmitter to that point and back to the receiver:

$$s(x',y') = \int \int f(x,y,z) e^{-j2k\sqrt{(x-x')^2 + (y-y')^2 + z_0^2}}\, dx dy. \tag{3.9}$$

Here, the object is assumed to be flat and parallel to the scanned aperture plane, i.e. at constant z. The wave number is denoted by $k = \omega/c$, where ω is the temporal angular frequency and c is the speed of light. The amplitude decay with range is not considered since it has little impact on the signal dependence on x' and y'. The exponential term in (3.9) approximates a spherical wave emanating from (x', y') with an effective wave number of $2k$. It can be decomposed into a superposition of plane-wave components:

$$e^{-j2k\sqrt{(x-x')^2 + (y-y')^2 + z_0^2}} = \int \int e^{jk_{x'}(x'-x) + jk_{y'}(y'-y) + jk_z z_0}\, dk_{x'} dk_{y'} \tag{3.10}$$

where $k_{x'}$ and $k_{y'}$ are the FT variables corresponding to x' and y', respectively. The spatial wave numbers $k_{x'}$ and $k_{y'}$ range from $-2k$ to $2k$ for the propagating waves. Using this relation, (3.9) is written as

$$s(x',y') = \int \int \left[\int \int f(x,y,z_0) e^{-j\left(k_{x'}x + k_{y'}y\right)}\, dx dy \right] \times e^{j\left(k_{x'}x' + k_{y'}y' + k_z z_0\right)}\, dk_{x'} dk_{y'}. \tag{3.11}$$

Assuming that the coordinate systems coincide, the distinction between the primed and unprimed coordinate systems is now dropped. The term in the brackets in (3.11) is the 2D FT of the reflectivity function $f(x, y, z_0)$. Thus, this expression can be written as

$$s(x,y) = \int\int \tilde{F}(k_x, k_y, z_0) e^{jk_z z_0} e^{j(k_x x + k_y y)} dk_x dk_y. \tag{3.12}$$

Thus, from (3.12) we can write

$$s(x,y) = FT_{2D}^{-1}\{\tilde{F}(k_x, k_y, z_0) e^{jk_z z_0}\} \tag{3.13}$$

where FT_{2D}^{-1} denotes the inverse 2D FT. By taking 2D FT of both sides of (3.13) we obtain

$$\tilde{F}(k_x, k_y, z_0) = FT_{2D}\{s(x,y)\} e^{-jk_z z_0}. \tag{3.14}$$

Thus, the object's reflectivity function (2D image) is reconstructed as

$$f(x,y,z_0) = FT_{2D}^{-1}\{FT_{2D}\{s(x,y)\} e^{-jk_z z_0}\}. \tag{3.15}$$

From the dispersion relation for the electromagnetic plane waves we have

$$k_x^2 + k_y^2 + k_z^2 = (2k)^2. \tag{3.16}$$

Thus, the Fourier variable k_z is determined as

$$k_z = \sqrt{4k^2 - k_x^2 - k_y^2}. \tag{3.17}$$

Finally, the reconstruction is summarized by the formula

$$f(x,y,z_0) = FT_{2D}^{-1}\{FT_{2D}\{s(x,y)\} e^{-j\sqrt{4k^2 - k_x^2 - k_y^2} \cdot z_0}\}. \tag{3.18}$$

The implementation of this single-frequency imaging system using RF circuitry has been proposed in [5] as shown in Figure 3.2. The signal is generated using a millimeter-wave oscillator, typically a Gunn diode oscillator. The signal is then transmitted using a small size wide-beamwidth antenna such as a small pyramidal waveguide horn. The illuminating waves are back-scattered and received by a receiver antenna. The receiver antenna is usually of the same size and type as the transmitting one and is adjacent to it. In this way, the pair of antennas can be considered as a single antenna, but with significantly higher transmitter to receiver isolation. The received signal is then mixed with in-phase (0° phase shift) and quadrature (90° phase shift) signals coupled from the millimeter-wave oscillator. The resulting signals are called I and Q representing the in-phase and quadrature components. These compose the complex scattered waves that are related to the amplitude and phase of the backscattered waves as

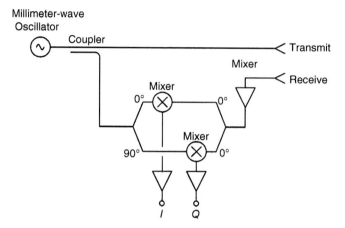

Figure 3.2 Simplified circuit schematic of the transceiver for the 2D holographic imaging [5].

$$I + jQ = Ae^{-j2kR} \tag{3.19}$$

where A is the amplitude of the scattered signal, R is the distance to the object, and k is the wave number. This signal is sampled over the 2D aperture using a 2D scanner to provide the required data for the image reconstruction algorithm $s(x, y)$.

3.3.2 Wideband Holographic 3D Imaging with Data Collected over Rectangular Apertures

In this section, we discuss wideband holographic imaging which is a means of forming 3D images of objects from wideband data gathered over a 2D aperture. The object is illuminated over a planar aperture similar to the technique discussed in Section 3.3.1 but a wideband illuminating source is employed instead of a single-frequency source. The reflected signal is then recorded coherently by the receiver, digitized, and stored in the computer. The data are mathematically processed to form a 3D image of the object's reflectivity function. This section details the algorithm presented in [5] that performs such image reconstruction. This technique relies extensively on the use of FTs, which may be computed very efficiently using the FFT algorithm. Similar to the 2D imaging technique discussed in Section 3.3.1, the limitation of the image resolution obtained with this technique stems from the well-known diffraction limit imposed by the wavelength as well as the source and receiver bandwidths, the size of the aperture, and the distance to the object. This method is similar to the single-frequency holography described in Section 3.3.1 with the extension to wideband illumination, which allows for 3D high-resolution imagery.

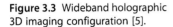
Figure 3.3 Wideband holographic 3D imaging configuration [5].

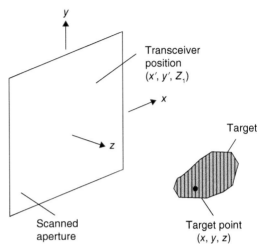

The image reconstruction technique assumes that the data represent a single reflection from the object (no multiple reflections), that there is no dispersion in the reflectivity function, and that there are no polarization changes due to the objects. The single reflection assumption is directly related to the well-known Born approximation [86, 123] and leads to a linearized model of scattering, which is discussed shortly.

Here, data collection is performed by scanning a transmitting source and a receiver over a rectilinear planar aperture that has one or more objects within its field of view. This system is quasi-monostatic, which means that the transmitting and receiving antennas are separate, but in approximately the same location, and may be assumed to be coincident at the midpoint between the two antennas.

The measurement configuration is shown in Figure 3.3, where the primed coordinates represent the position of the transceiver and the unprimed coordinates represent a point in the object or image space. If the object is characterized by a reflectivity function $f(x, y, z)$, then the response measured by the transceiver can be approximated as a superposition of spherical waves, the strength of which scales with the reflectivity function whereas their phase delay is determined by the round-trip signal path to the object. If the measurement plane is at $z = Z_1$, the round-trip phase delay is

$$2k\sqrt{(x-x')^2 + (y-y')^2 + (z-Z_1)^2} \tag{3.20}$$

where $k = \omega/c$ is the wave number, ω is the temporal angular frequency, and c is the speed of light.

The response measured at the transceiver is

$$s(x',y',\omega) = \int\int\int f(x,y,z) \cdot e^{-j2k\sqrt{(x-x')^2+(y-y')^2+(z-Z_1)^2}} dxdydz. \qquad (3.21)$$

Here, again the amplitude decay with distance is not considered since it has little impact on the signal dependence on (x', y'). Alternatively, the data could be collected in the time domain. In this case, the data in the form of (3.21) is obtained by applying FT on the recorded time-domain signals $s_t(x', y', t)$:

$$s(x',y',\omega) = FT_{(t)}[s_t(x',y',t)] \qquad (3.22)$$

where $FT_{(t)}$ indicates FT operation with respect to time.

The exponential term in (3.21) represents a spherical wave, which can be decomposed into an infinite superposition of plane waves:

$$e^{-j2k\sqrt{(x-x')^2+(y-y')^2+(z-Z_1)^2}} = \int\int e^{-jk_{x'}(x-x')-jk_{y'}(y-y')-jk_z(z-Z_1)} dk_{x'}dk_{y'}$$

$$(3.23)$$

where $k_{x'}$ and $k_{y'}$ are the FT variables corresponding to x' and y', respectively. The spatial wave numbers $k_{x'}$ and $k_{y'}$ range from $-2k$ to $2k$ for propagating waves. This indicates that this imaging strategy assumes far-field measurements, where only the propagating (non-evanescent) components of the scattered waves are acquired.

Using the decomposition into plane waves and rearranging yields

$$s(x',y',\omega) = \int\int\left[\int\int\int f(x,y,z) \cdot e^{-j(k_{x'}x+k_{y'}y+k_zz)} dxdydz\right] \cdot e^{jk_zZ_1} e^{jk_{x'}x'} e^{jk_{y'}y'} dk_{x'}dk_{y'}$$

$$(3.24)$$

where the triple integral in (3.24) represents the 3D FT of the reflectivity function. Therefore, (3.24) can be written as

$$s(x',y',\omega) = \int\int \tilde{F}(k_{x'},k_{y'},k_z) e^{jk_zZ_1} e^{j(k_{x'}x'+k_{y'}y')} dk_{x'}dk_{y'} = FT_{2D}^{-1}\{\tilde{F}(k_{x'},k_{y'},k_z) e^{jk_zZ_1}\}.$$

$$(3.25)$$

Taking the 2D FT of both sides of (3.25) and dropping the distinction between the primed and unprimed coordinate systems yields

$$FT_{2D}\{s(x,y,\omega)\} \equiv \tilde{S}(k_x,k_y,\omega) = \tilde{F}(k_x,k_y,k_z) e^{jk_zZ_1}. \qquad (3.26)$$

In order to use (3.26) for image reconstruction, the frequency ω needs to be expressed as a function of k_z. This is accomplished using the dispersion relation for plane waves in free space:

$$k_x^2 + k_y^2 + k_z^2 = (2k)^2 = 4\left(\frac{\omega}{c}\right)^2. \qquad (3.27)$$

Using this relation and inverting (3.26) yields

$$f(x,y,z) = \text{FT}_{3D}^{-1}\left\{\tilde{F}(k_x,k_y,k_z)\right\} \tag{3.28}$$

where

$$\tilde{F}(k_x,k_y,k_z) = \tilde{S}(k_x,k_y,\omega)e^{-jk_z Z_1} \tag{3.29}$$

and

$$k_z = \sqrt{4k^2 - k_x^2 - k_y^2} = \sqrt{4\left(\frac{\omega}{c}\right)^2 - k_x^2 - k_y^2}. \tag{3.30}$$

Combining the above relations, results in the reconstruction of the reflectivity function (3D image) as

$$f(x,y,z) = \text{FT}_{3D}^{-1}\left\{\text{FT}_{2D}\{s(x,y,\omega)\}\cdot e^{-j\sqrt{4k^2 - k_x^2 - k_y^2}Z_1}\right\}. \tag{3.31}$$

Equation (3.31) can be employed directly to reconstruct a 3D image if the data are defined continuously in x, y, and ω. However, in practice, the data $s(x, y, \omega)$ are discretely sampled at uniform intervals of position and temporal frequency. Since the data are uniformly sampled in x and y, the 2D FFT may be used to obtain a sampled version of $\tilde{S}(k_x,k_y,\omega)$. But the angular frequency ω is a function of k_x, k_y, and k_z, as shown by (3.30). Therefore, $\tilde{S}(k_x,k_y,\omega)$ is a mapping of $\tilde{S}(k_x,k_y,k_z)$, where the samples along the k_z axis are non-uniformly spaced. In the 3D (k_x, k_y, k_z) space, the uniformly sampled $\tilde{S}(k_x,k_y,\omega)$ function yield points, which are equispaced along k_x and k_y but lie on concentric spheres of radius $2k$. In order to perform the inverse 3D FFT in (3.28), the data need to be resampled to uniformly spaced positions in k_z. This is accomplished using linear interpolation techniques.

3.3.2.1 Spatial and Frequency Sampling

In order to collect sufficient data for a successful image reconstruction process, Nyquist sampling criterion need to be satisfied. The spatial sampling along the aperture is determined by a number of factors including the wavelength, the size of the aperture, the size of the object, and the distance to the object. To satisfy the Nyquist criterion the phase shift from one sample point to the next one should be less than π rad. The largest phase shift occurs for an object very close to the aperture and the sample points near the edge of the aperture. For a spatial sampling interval of Δx, the worst case must have a phase shift of no more than $2k\Delta x$. Therefore, the sampling criterion can be expressed as

$$\Delta x < \frac{\lambda}{4} \tag{3.32}$$

where λ is the wavelength. This limit is more restrictive than what is usually required since in many image reconstruction experiments, the object is a moderate distance from the aperture and the antenna beamwidth is usually less than 180°. Thus, holographic imaging systems can often employ sampling intervals on the order of $\lambda/2$ and still provide high-quality images.

The frequency sampling Δf is determined in a similar way using Nyquist sampling criterion. The maximum phase shift between two consecutive frequencies occurs for signals reflected by the most remote targets. The position such targets determines the maximum range R_{max} of the imaging system. For a frequency sampling step of Δf a wave number sampling step of $\Delta k = 2\pi\Delta f/c$ is obtained, which in turn contributes to the maximum phase shift of $2\Delta k R_{max}$. Requiring that this phase shift be less than π rad yields

$$\Delta k < \frac{\pi}{2R_{max}} \tag{3.33}$$

or

$$\Delta f < \frac{c}{4R_{max}}. \tag{3.34}$$

From (3.34), the number of frequency samples for a bandwidth B is obtained as

$$N_f > \frac{2R_{max}}{(c/2B)}. \tag{3.35}$$

Equation (3.35) can be interpreted in terms of the range resolution $c/2B$ (as discussed in the following section). It shows that at least two frequency samples are needed per range resolution cell.

3.3.2.2 Range and Cross-Range Resolution

Since the basis of holographic image reconstruction is on the formation of the image in the spatial-frequency domain, the spatial resolution of the image is determined by the extent of spatial-frequency coverage. For simplicity let's consider one axis. Uniform frequency coverage (*rect* function) of width Δk results in a spatial resolution of

$$\delta = \frac{2\pi}{\Delta k}. \tag{3.36}$$

This is distance between the major axis and the first null of a *sinc* function resulted when taking the inverse FT of a rectangular spectrum with width Δk (for a brief review of resolution estimating, please see the Appendix). For the 3D image reconstruction, the spatial-frequency coverage has a ring shape as shown in Figure 3.4. If this region is approximated as rectangular, then the width in the k_x-direction is approximately $4k_c \sin(\theta_b/2)$, where k_c is the wave number at the center frequency and θ_b is the lesser of the full beamwidth of the antenna or

Figure 3.4 Spatial-frequency coverage in the range and cross-range directions [5].

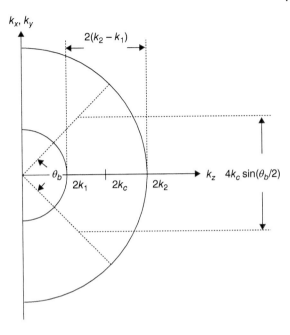

the angle subtended by the aperture. Thus, from (3.36), the cross-range resolution can be approximated by

$$\delta_x \approx \frac{\lambda_c}{4\sin\left(\dfrac{\theta_b}{2}\right)}. \tag{3.37}$$

For a system with the range R much greater than the aperture length D, we can write

$$\sin\left(\frac{\theta_b}{2}\right) \approx \frac{D}{2R} \tag{3.38}$$

And then the cross-range resolution can be approximated by

$$\delta_x \approx \frac{\lambda_c}{2} F\# \tag{3.39}$$

where $F\# = R/D$ is the optical F number. Similar expression can be derived for the cross-range resolution in the y-direction except that the F number or beam-width may be different.

From Figure 3.4, the spatial-frequency coverage along k_z is $2(k_2 - k_1)$, where k_1 and k_2 are the wave numbers at the low and high frequencies of the system (see (3.30) for the relation between the wave number and k_z). Using similar approach as discussed above, this spatial-frequency coverage results in a range resolution of approximately

$$\delta_z \approx \frac{2\pi}{2(k_2 - k_1)} = \frac{c}{2B} \tag{3.40}$$

where B is the temporal frequency bandwidth of the system.

3.3.2.3 Sample Experimental Images

A wideband holographic transceiver can be constructed similar to the single-frequency transceiver shown in Figure 3.2, except that the frequency is swept over a range of interest. A voltage-controlled oscillator (VCO) can be employed to generate the required millimeter-wave with the capability of sweeping the frequency. The signals must be sampled according to the criteria described in Section 3.3.2.1.

Figure 3.5a shows the optical image of a Glock-17 hand gun. The trigger guard and parts of the handle of the Glock are made from plastic. Figure 3.5b shows a 100–112-GHz wideband holographic image of this gun. This image has a resolution better than 3 mm. The image has high quality. The plastic components can be observed in the image but, as expected, have lower intensity. Also, Figure 3.6a shows the optical image of a plastic Olin flare gun. This gun is entirely plastic, except for a small piece of metal near the start of the barrel. Figure 3.6b shows the wideband holographic image for this gun. The metal component in the gun is clearly brighter in the image. To measure the scattered waves in both the examples in Figures 3.5 and 3.6, an aperture of approximately 30 cm × 30 cm has been synthesized with a discretization of $256(x) \times 256(y)$ and 128 frequency samples have been measured. The scan time has been

(a)   (b)

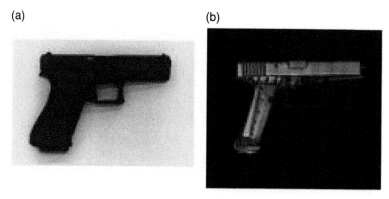

Figure 3.5 (a) Optical image and (b) 100–112 GHz wideband holographic image of a Glock-17 handgun [5].

(a) (b)

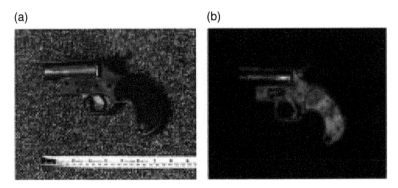

Figure 3.6 (a) Optical image and (b) 100–112 GHz wideband holographic image of an Olin plastic flare gun [5].

(a) (b)

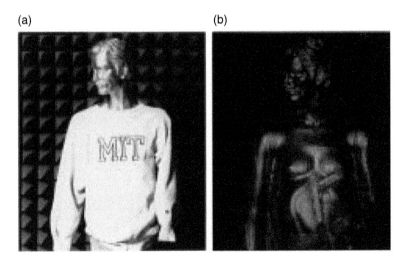

Figure 3.7 Optical image and (b) 100–112 GHz wideband holographic image of a clothed mannequin with a concealed Glock-17 handgun [5].

approximately 20 min. In a more complicated imaging experiment, a mannequin with a concealed weapon is imaged. Figure 3.7 shows optical and millimeter-wave (100–112 GHz) images related to this example. This example clearly demonstrates the capability of the millimeter-wave holographic imaging for security screening applications. It has been reported in [5] that fine details can be observed in the image including the Massachusetts Institute of Technology (MIT) logo and the duct tape used to attach the weapon to the mannequin. In the example in Figure 3.7, the scattered waves have been scanned over an aperture of approximately 80 cm × 140 cm with a discretization of $512(x) \times 1024(y)$ and 128 frequency samples have been measured. The scanning time has been reported as three hours.

In practical security screening system used in public places, the wideband holographic imaging of the full-body needs to be implemented in a few seconds. The sampling along the aperture needs to be implemented every half wavelength or so as discussed in Section 3.3.2.1 and the size of aperture need to be at least $0.70\,\text{m} \times 2.0\,\text{m}$. Since scanning a single transceiver with these sampling requirements takes a long time (at least several minutes), in [5], an electronically switched array of antennas has been employed to expedite the process. In an ideal scenario, a 2D array of antennas would provide sampling of the whole planar aperture in one shot without any mechanical motion. However, the cost and complexity of a large 2D high-resolution array renders it impractical. A linear array can, however, be fabricated at a reasonable cost and can be scanned quickly along the axis perpendicular to the array to gather data over full 2D aperture.

Using the concept described above, in [5], a wideband imaging system has been developed that utilizes linear array of antennas mechanically scanned over the axis perpendicular to the array to quickly synthesize the 2D aperture data. Figure 3.8 shows a photograph of this system. Also, Figure 3.9 shows a block diagram of the system. The system consists of a sequentially switched linear array of antennas driven by a wideband millimeter-wave transceiver. The array and transceiver are mounted on a large fast mechanical scanner, which allows data acquisition over a 2D aperture in less than one second. During a scan, the array is sequenced electronically at very high speed to gather each line of data as

Figure 3.8 Wideband holographic millimeter-wave imaging system in [5].

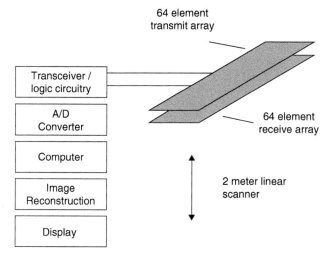

Figure 3.9 Block diagram of a holographic security screening system in [5].

the scanner is moving along the other axis, thereby gathering data over a full 2D aperture. This system operates in the Ka-frequency band (26.5–40 GHz). There are 128 antenna elements organized as an upper row of 64 transmit antennas and a lower row of 64 receive antennas. Logic circuitry sequences the transmitting and receiving antennas to transmit from one antenna and receive the wideband signal from each of the two neighboring antennas in the receive row. This places a virtual sampling point halfway in between each transmit and receive antenna.

Since in the system in Figure 3.9 the transmit row and receive row are offset by half the antenna spacing, thus, the effective sample spacing is one-half of the single-row antenna spacing. This sequencing scheme cannot be used on the last antenna element, thus, the effective number of sample points is 127. The horizontal aperture width is 72.6 cm, which yields an effective sample spacing of 5.7 mm.

Using this scanning methodology, the wideband signal is sampled horizontally across the array and vertically over the aperture. It is digitized by an analog-to-digital (A/D) converter for subsequent storage and processing in the computer. After digitizing, the reconstruction algorithm is applied to reconstruct the 3D image of the object. Using image processing techniques, the 3D image is then collapsed into a fully focused 2D image of the object for display on the computer.

A photograph of the wideband array is shown in Figure 3.10. The transmit-row switched antenna array is composed of eight single-pole eight-throw (SP8T) p-i-n-diode waveguide switches, interconnected with a waveguide

Figure 3.10 Photograph of the linear array [5].

manifold feed and fed by a ninth SP8T waveguide switch. Each of the 64 outputs has a small dielectric polyrod antenna inserted into the waveguide. The polyrods can be of any length, however, the ones used here are four wavelengths long (at 35 GHz). The receive-row switched antenna array is identical and is placed back to back with the transmit array. Integrated switch drivers are contained within each SP8T switch module, allowing a simple connector with power and coded logic inputs to each of the 18 SP8T switch modules.

A simplified schematic of circuit of the transceiver is shown in Figure 3.11. The RF VCO is swept from 27 to 33 GHz. This transceiver is a heterodyne design with the heterodyne offset frequency obtained by offsetting the local oscillator (LO) by an IF (in this case, 600 MHz). An IF reference signal (at 600 MHz) is obtained by mixing coupled signals from the RF and LO oscillators.

The phase and amplitude of the scattered signal are contained within the IF receive signal, which is the received signal down-converted using the LO oscillator. The in-phase and quadrature signals are obtained by mixing the two signals with and without a 90° phase shift, as shown in the figure.

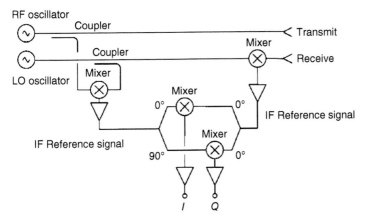

Figure 3.11 Simplified schematic of a wideband millimeter-wave transceiver [5].

The I and Q signals contain the complex scattered wave's amplitude and phase $I + jQ = Ae^{-j2kR}$, where A is the amplitude of the scattered wave and R is the range to the object.

This signal is sampled over the frequency range of the system (27–33 GHz) and sampled over an x–y aperture using the sequentially switched linear array and vertical scanner to obtain the input to the image reconstruction algorithm $s(x, y, \omega)$.

The signal is discretely sampled with typical dimensions of 127 samples, 512 samples, and 64 frequency samples with spatial and frequency steps sizes of 5.7 mm, 3.9 mm, and 94 MHz, respectively. The frequency sweep time is 20 seconds and the integration time per frequency is 0.3 second. The IF bandwidth is approximately 2.5 MHz. The image reconstruction time is very dependent on the type of computer system used to perform the image reconstruction. The Unix-based SUN computers (approximately 1995) required several minutes to perform the reconstruction. It has been reported in [5] that a dedicated multiprocessor digital signal processor (DSP) (Analog Devices SHARC-based, 1998) requires less than 10 seconds to perform the reconstruction.

Figure 3.12 shows the significant improvement in the image quality that has been obtained by converting the single-frequency millimeter-wave imaging system to wideband operation. The single-frequency images show significant degradation due to lack of focus over many parts of the image. In addition, some degradation is apparent due to poor sensitivity in the single-frequency transceiver. By contrast, the wideband images are fully focused due to the 3D image reconstruction. Significantly higher dynamic range is also apparent in the wideband images. In [5], this has been reported due to the better focus as well as the higher sensitivity of the transceiver design.

(a)   (b)   (c)   (d)

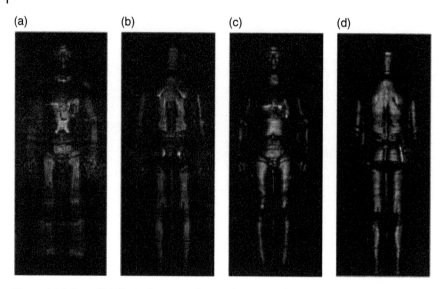

Figure 3.12 (a and b) Single-frequency (35 GHz) images of a man. (c and d) Wideband (27–33 GHz) images of the same man [5].

3.3.3 Wideband Holographic 3D Imaging with Data Collected over Cylindrical Apertures

In a cylindrical imaging system, the inspected medium is illuminated from all possible angles leading to higher quality images. Data acquisition can be performed by scanning the transmitter and receiver antennas up to 360° around an object near the axis of rotation, as illustrated in Figure 3.13 [11].

Similar to the planar 3D imaging technique discussed in Section 3.3.2, the cylindrical image reconstruction algorithm derived in this section is an extension of SAR imaging in [122] where a reconstruction process has been derived for a 1D circular "aperture" resulting in a 2D image in the plane of the circle. In [11], this technique was extended by adding the vertical axis, which makes the resulting image 3D. The details of the derivation of the holographic image-reconstruction algorithm for data collected over cylindrical apertures have been presented in a US patent [124] and also in [125]. Below, the reconstruction algorithm is summarized.

3.3.3.1 Image Reconstruction Technique

For a cylindrical scan of radius R configured as shown in Figure 3.13, the transceiver position is $(R \cos \theta', R \sin \theta', z')$. The object is assumed to be near the z axis with an arbitrary point on the object at (x, y, z). The round-trip phase delay from the transceiver to the object point is

$$2k\sqrt{(R\cos\theta' - x)^2 + (R\sin\theta' - y)^2 + (z' - z)^2}. \tag{3.41}$$

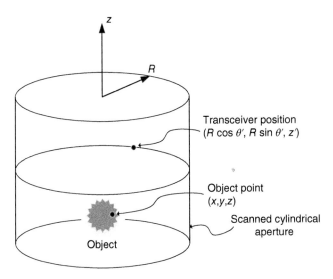

Figure 3.13 Cylindrical holographic imaging configuration [11].

The signal received at (θ', ω, z') can be expressed as the superposition of differential backscattering waves arising at each point in the object, where the reflectivity function is $f(x, y, z)$ and the round-trip phase delay is defined in (3.41), i.e.

$$s(\theta', \omega, z') = \iiint f(x, y, z) \times e^{-j2k\sqrt{(R\cos\theta' - x)^2 + (R\sin\theta' - y)^2 + (z' - z)^2}}\, dxdydz.$$

(3.42)

By denoting the transceiver position in Cartesian coordinate system as (x', y', z'), (3.42) can be written as

$$s(x', y', z') = s(\theta', \omega, z') = \iiint f(x, y, z) \times e^{-j2k\sqrt{(x' - x)^2 + (y' - y)^2 + (z' - z)^2}}\, dxdydz.$$

(3.43)

Here, we employ the spectral domain decomposition of the free space Green's function written as [126]

$$g(r, r') = \frac{e^{-jk(r' - r)}}{4\pi |r' - r|} = \frac{-1}{(2\pi)^3} \iiint_k \frac{e^{-j[k_x(x' - x) + k_y(y' - y) + k_z(z' - z)]}}{k^2 - k_x^2 - k_y^2 - k_z^2}\, dk_x dk_y dk_z.$$

(3.44)

From (3.43) and (3.44), and after neglecting the amplitude terms in the spatial and spectral domains, the signal measured by the transceiver can be written as

$$s(\theta', \omega, z') = \frac{-1}{(2\pi)^3} \iiint f(x, y, z) \iiint_k e^{-j[k_x(x' - x) + k_y(y' - y) + k_z(z' - z)]}\, dk_x dk_y dk_z dxdydz.$$

(3.45)

The Cartesian spectral Fourier variables can be substituted by their corresponding variables in the cylindrical coordinate system (k_r, θ, k_z) using

$$k_x = k_r \cos\theta \tag{3.46}$$

$$k_y = k_r \sin\theta \tag{3.47}$$

where

$$k_r = \sqrt{k_x^2 + k_y^2} \tag{3.48}$$

and

$$\theta = \arctan\left(\frac{k_y}{k_x}\right). \tag{3.49}$$

After substituting and rearranging we obtain

$$s(\theta', \omega, z') = \iiint\left[\iint\int f(x,y,z)e^{j[k_x x + k_y y + k_z z]}dxdydz\right]e^{-j[k_r \cos\theta x' + k_r \sin\theta y' + k_z z']}k_r dk_r d\theta dk_z. \tag{3.50}$$

The triple integral in the brackets in (3.50) can be considered as the FT of the contrast function for the object $\tilde{F}(k_x, k_y, k_z)$. Thus, (3.50) can be written as

$$s(\theta', \omega, z') = \iint\left[\int\tilde{F}(k_x, k_y, k_z)e^{-ik_z z'}dk_z\right]e^{-j[k_r \cos\theta x' + k_r \sin\theta y']}k_r dk_r d\theta. \tag{3.51}$$

Again the term in the brackets in (3.51) can be considered as the inverse FT of $\tilde{F}(k_x, k_y, k_z)$ with respect to the k_z variable. Thus, taking FT of both sides of (3.51) with respect to the z' variable leads to

$$s(\theta', \omega, k_z) = \iint\tilde{F}(k_x, k_y, k_z)e^{-j[k_r \cos\theta x' + k_r \sin\theta y']}k_r dk_r d\theta. \tag{3.52}$$

Writing (3.52) only in terms of Fourier variables in the cylindrical coordinate system leads to

$$s(\theta', \omega, k_z) = \int\left[\int\tilde{F}(k_r, \theta, k_z)\cdot e^{-j[k_r R\cos(\theta - \theta')]}d\theta\right]k_r dk_r. \tag{3.53}$$

The integral inside the brackets in (3.53) can be interpreted as a convolution integral along θ variable. Thus, taking FT with respect to the θ variable leads to

$$s(k_\theta, \omega, k_z) = \int\tilde{F}(k_r, k_\theta, k_z)\cdot F_\theta\left\{e^{-j[k_r R\cos(\theta)]}\right\}k_r dk_r. \tag{3.54}$$

where F_θ denotes the FT operation with respect to the θ variable. The integration variable in (3.54) can be further changed to k_r^2. Under a far-field assumption, this variable in the spectrum for the object is constrained to values belonging to the Ewald sphere only:

$$4k^2 = k_r^2 + k_z^2. \tag{3.55}$$

This imposes a δ-function behavior of the integrand in (3.54) with respect to k_r^2. Thus the object's contrast function reconstruction in the spectral domain can be implemented as

$$\tilde{F}(k_r,\theta,k_z) = F_{k_\theta}^{-1}\left\{\frac{s(k_\theta,\omega,k_z)}{F_\theta\{e^{-j[k_r R\cos\theta]}\}}\right\} \tag{3.56}$$

where k_r must satisfy (3.55). In order to reconstruct the object in the Cartesian domain, first the reconstructed function in (3.56) is interpolated into a Cartesian spectral grid. This operation is called *Stolt mapping*, which realizes the transformation from the cylindrical Fourier space into the Cartesian Fourier space using (3.46)–(3.48) and (3.55), i.e.

$$[k_x,k_y,k_z]^T = [k_r\cos\theta, k_r\sin\theta, k_z]^T, k_r = \sqrt{4k^2 - k_z^2}. \tag{3.57}$$

Thus, the final inversion process can be summarized as

$$f(x,y,z) = F_{k_x,k_y,k_z}^{-1}\left\{\text{Stolt mapping}\left[F_{k_\theta}^{-1}\left\{\frac{s(k_\theta,\omega,k_z)}{F_\theta\{e^{-jk_r R\cos\theta}\}}\right\}\right]\right\}. \tag{3.58}$$

The implementation of (3.58) necessitates the numerical implementation of the function $e^{-jk_r R\cos\theta}$ and its subsequent Fourier transformation. However, as proposed in [11, 124], the term in the denominator can be approximated by an analytical expression valid for $k_\theta \ll 2k_r R$ (object close to the z axis), which leads to a fully analytical inversion formula:

$$f(x,y,z) = F_{k_x,k_y,k_z}^{-1}\left\{\text{Stolt mapping}\left[F_{k_\theta}^{-1}\left\{s(k_\theta,\omega,k_z)\cdot e^{-j\sqrt{4k_r^2 R^2 - k_\theta^2}}\right\}\right]\right\}. \tag{3.59}$$

3.3.3.2 Sampling Criteria and Spatial Resolution

As with the rectilinear method, this technique relies extensively on the use of FTs, which may be computed very efficiently using the FFT algorithm. Therefore, the only limitation on the resolution of the reconstructed images is the diffraction-limited resolution imposed by the wavelength of operation and the source and receiver beamwidths. The image is reconstructed in Cartesian coordinate system which is convenient for display using conventional computer displays.

The sampling requirements for a practical cylindrical imaging system are similar to those for the rectilinear imaging system. The sampling step along the vertical axis must be smaller than quarter-wavelength to satisfy the Nyquist criterion. As discussed in Section 3.3.2.1, in practice, half-wavelength is usually adequate for many scenarios using moderate beamwidth antennas and objects located away from the aperture. In the angular direction, the Nyquist criterion will be satisfied if the phase shift from one sample point to the next is less than π rad. The worst-case phase shift for an object contained by $r \leq R_{obj}$, where R_{obj} is the maximum radius of the object, results in a required sampling interval of

$$\Delta\theta < \frac{\pi}{2kR_{obj}}. \tag{3.60}$$

The sampling required for the frequency sweep is identical to that of rectilinear imaging technique discussed in Section 3.3.2.1, with a required frequency sample interval of $\Delta f < c/(2R_{max})$, where R_{max} is the maximum object range.

Similar to the rectilinear system, the lateral resolution in the vertical direction is given by

$$\delta_z \approx \frac{\lambda_c}{4\sin\left(\dfrac{\theta_b}{2}\right)} \tag{3.61}$$

where the angle θ_b, is the lesser of the full vertical beamwidth of the antenna or the angle subtended by the vertical aperture. The lateral resolution in the azimuthal direction for the cylindrical imaging technique is given by [11]

$$\delta_\theta \approx \frac{\lambda_c}{4\sin\left(\dfrac{\theta_b}{2}\right)} \tag{3.62}$$

where in this case, θ_b will be the angle subtended by the cylindrical aperture. This angle is commonly restricted to angles of less than 180°.

3.3.3.3 Image Reconstruction Results

The cylindrical imaging technique developed at PNNL operates using a vertically oriented linear array that sweeps a cylindrical aperture using a mechanical scanner [124]. The primary advantage of the cylindrical imaging technique over the rectilinear technique is that the object can be "viewed" from multiple angles. In the system developed at PNNL, a sequentially switched linear antenna array electronically sequences along the vertical aperture at high speed (typically less than 1 ms), and the cylindrical mechanical scan is then used to sample the cylindrical aperture along the azimuthal direction (typically in 1–10 seconds). The use of electronically switched arrays is critical for practical application of this cylindrical imaging system. This allows high-speed data acquisition and near-real-time imaging performance for example for security screening applications.

The amplitude and phase of the scattered wave front at each aperture position are sampled using the millimeter-wave transceiver, which transmits a beam and sweeps over a wide frequency bandwidth. This 3D (angle, z, frequency) dataset is then mathematically reconstructed using the algorithm described above to form fully focused, diffraction limited, 3D images of the object or person under surveillance.

The reconstructed images show differences in the reflectivity, orientation, and shape of the object. In [11], 8–64 evenly spaced overlapping 90° segments of the data are reconstructed to form a video animation in which the image appears to rotate. This allows convenient and rapid inspection of the imaging results by a screener. Also, a "combined" imaging technique has been developed that combines the 3D reconstruction results from eight overlapping 90° reconstructions to form a single 3D reconstruction with all 360° of angular illumination represented in the dataset. This 3D image is then rendered using computer graphics visualization techniques.

The performance of the millimeter-wave cylindrical imaging technique is demonstrated via an example shown in Figure 3.14, for a mannequin carrying a number of concealed items [11]. The concealed items include various plastic bottles, glass vials, plastic explosive simulant, and a handgun. The cylindrical data are collected using a laboratory scanner and a 40–60 GHz radar imaging transceiver. The expected lateral resolution is approximately half-wavelength (3 mm) whereas the depth resolution is approximately 7.5 mm. A total of 64 overlapping 90° arc-segment image reconstructions were performed, each resulting in a 3D image centered on varying aspect angles. Each frame shown has been reduce from a 3D image to a 2D image by image processing techniques to make it suitable for display. Visible threats or objects in the images include a handgun (lower left thigh), explosive simulant (right thigh), and plastic and glass vials/bottles (front and back torso). Details of clothing seams and buttons are also clearly resolved in the images.

Figure 3.15 shows a front-view projection image from the same cylindrical dataset but using the combined image reconstruction. The illumination is much more extensive in this image, due to the inclusion of the full 360° dataset, rather than the 90° segment used to reconstruct the images in Figure 3.14.

Polarimetric imaging techniques can be employed to obtain additional information about the object. Circular polarization is particularly interesting because circularly polarized waves incident on relatively smooth reflecting objects are typically reversed in their rotational handedness, e.g. left-hand circular polarization (LCP) is reflected to become right-hand circular polarization (RCP). An incident wave that is reflected twice (or any even number of times) prior to returning to the transceiver has its handedness preserved. Sharp features, such as wires and edges, tend to return linear polarization, which can be considered to be a sum of both LCP and RCP. These characteristics can be exploited for

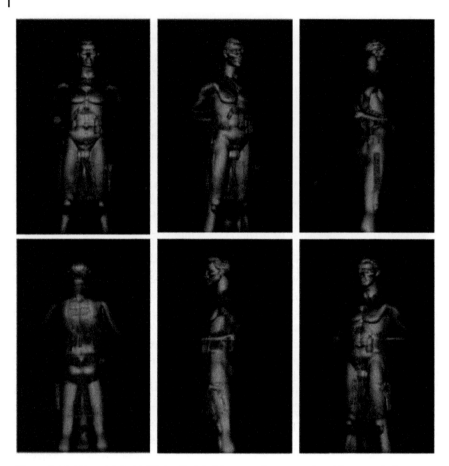

Figure 3.14 U-band (40–60 GHz) cylindrical images of mannequin with concealed weapons and items. *Source:* Reprinted with permission from [11], OSA.

personnel screening by allowing differentiation of smooth features, such as the body, from sharper features present in many concealed items.

In [11], a laboratory imaging system was set up to explore the characteristics of the circular polarization imaging system and obtain imaging results. The experimental imaging configuration used a rotating platform placed in front of a rectilinear (x–y) scanner. This system emulates a linear-array-based cylindrical imaging system by mechanically scanning the transceiver at each rotational angle of the platform. The system was set up to operate over the 10–20 GHz frequency range. Imaging results from a clothed mannequin carrying a concealed handgun and simulated plastic explosives are shown in Figure 3.16. Images were obtained using 90° arc segments of cylindrical data centered at 64 uniformly spaced angles, ranging from 0° to 360°, with sample images shown at approximately 30° and 180° in the figure.

Figure 3.15 U-band (40–60 GHz) combined cylindrical image of mannequin with concealed weapons and items. *Source:* Reprinted with permission from [11], OSA.

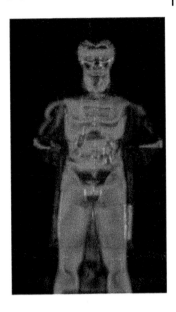

Three polarization combinations were imaged using otherwise identical experimental parameters. HH refers to transmitting and receiving with horizontal electric field polarization. RL refers to transmitting RCP and receiving LCP. RR refers to transmitting RCP and receiving RCP. The HH images are shown on the left side of Figure 3.16, RL images in the center, and RR images on the right. An interesting feature of these images is that the polarization properties can highlight multipath signal returns. In the HH images, multipath artifacts can be observed between the thighs and between the mannequin's right arm and body. These artifacts are not present in the RL image, which suppresses double (or even) bounce reflections.

The RR image highlights, or isolates, the double-bounce reflections and suppresses the single bounce return. Similar results are observed in the back view images in Figure 3.16. The primary multipath artifacts in the HH image are between the upper arm and the body of the mannequin. These artifacts are eliminated in the RL image and isolated in the RR image. The concealed weapons are enhanced in the RR images. The edges of the concealed handgun are highlighted in the RR images due to the dihedral (double bounce) reflection formed around the perimeter of the handgun as placed on the body of the mannequin. Similarly, the edges of the simulated plastic explosive are highlighted in the RR image of the back of the mannequin. These polarization properties in the images can be exploited to enhance detection of concealed objects and reduce privacy concerns.

The PNNL wideband holographic millimeter-wave imaging technology has been licensed to L3-SafeView, (website: http://www.safeviewinc.com). L3-SafeView has produced a commercial system, the ProVision Whole Body Imager. The ProVision

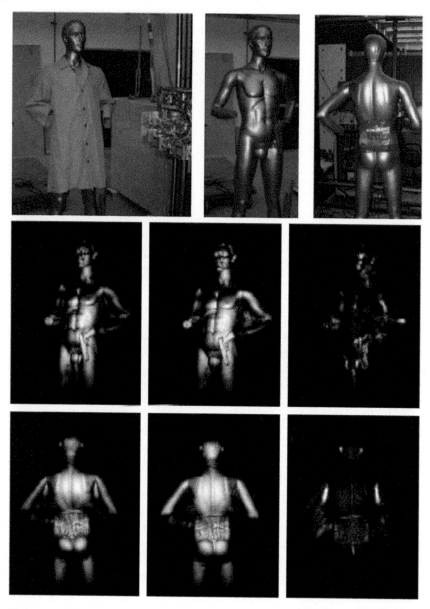

Figure 3.16 Polarimetric imaging results at 10–20 GHz for a mannequin with concealed metal gun on its abdomen and simulated plastic explosive on its lower back. Left-side images are HH polarization, center images are RL polarization, and right-side images are RR polarization. *Source:* Reprinted with permission from [11], OSA.

operates within the 24.25–30 GHz band and incorporates two vertical linear antenna arrays with 384 antenna elements in each array. The ProVision configuration has an open entry and exit of the scanner and allows convenient high-throughput operation while still providing complete imaging of the subject. This system can acquire a scan in approximately 1.5 seconds and presents images to the operator within 2 seconds. ProVision systems are currently deployed around the world at airports, courthouses, and other governmental buildings, military checkpoints, and commercial buildings.

4

Microwave/Millimeter-Wave Holography for Near-Field Imaging Applications

In this chapter, we discuss recent developments of microwave holography toward near-field applications such as biomedical imaging, nondestructive testing, and underground imaging.

The holographic methods presented in [5, 11] rely on an assumed analytical (exponential) form of the incident field and Green's function in order to cast the inversion in the form of a three-dimensional (3D) inverse Fourier transform (FT). This assumption is sufficiently accurate for applications such as millimeter-wave concealed weapon detection leading to high-resolution images. This is due to the fact that the electrical distance between the antennas and the imaged body is sufficiently large to allow for employing far-field approximations of the illuminating and scattered waves. However, for applications such as microwave biomedical imaging or nondestructive testing of materials, the far-field analytical approximations are inadequate leading to low quality of the reconstructed images [86]. To improve the accuracy of the holographic imaging techniques in near-field applications, two major approaches have been presented based on: (i) using the simulated incident field and Green's function information and (ii) using the measured point-spread function (PSF) of the imaging system. Similar to the techniques in Chapter 3, these techniques utilize phase and amplitude information recorded over a two-dimensional (2D) aperture and perform digital image reconstruction. However, they do not impose far-field approximations on the waves (plane or spherical waves). In this chapter, these techniques are described in detail.

4.1 2D Near-Field Holographic Imaging

In [127], the single-frequency 2D holographic image reconstruction developed in [5] is adapted for near-field imaging applications. This method does not make any assumptions about the incident field such as those based on spherical or

Real-Time Three-Dimensional Imaging of Dielectric Bodies Using Microwave/Millimeter-Wave Holography, First Edition. Reza K. Amineh, Natalia K. Nikolova, and Maryam Ravan.
© 2019 by The Institute of Electrical and Electronics Engineers, Inc.
Published 2019 by John Wiley & Sons, Inc.

cylindrical waves. The incident field can be even given in a numerical or discretized form as in the case when it is derived through electromagnetic simulation or measurement. This is especially important in near-field imaging where the object is close to the antenna and the spherical assumption for the illuminating wave is not valid. This also makes this technique applicable to layered background media. Also, in contrast to the original microwave holography, the method in [127] can be applied to both transmitted and reflected signals.

In [127], the *S*-parameters at the terminals of two antennas (one transmitting and one receiving) are measured when the antennas scan together two separate parallel apertures on both sides of a 2D object (assuming the thickness of the object along the range direction is negligible). The *S*-parameters are then processed to first localize the object in the range direction and then reconstruct a 2D image of the object. Unlike the 2D holographic method in [5], this method does not need *a priori* knowledge of the object location along the range axis.

Figure 4.1 shows the acquisition setup consisting of a transmitting antenna, a receiving antenna, and a 2D object between the antennas. The transmitting and the receiving antennas perform a 2D scan while moving together on two separate planes (transmitter and receiver planes) positioned at $z' = 0$ and $z' = D$, respectively. The object is positioned at $z = \bar{z}$ and its thickness along the *z*-axis is assumed to be negligible. Assume we know the incident field at $z = \bar{z}$ as a function of *x* and *y* when the transmitting antenna is at $x' = 0$, $y' = 0$ ($z' = 0$ at the transmitting plane):

$$E_0^{\text{inc}}(x,y,\bar{z}) \equiv E^{\text{inc}}(x,y,\bar{z};0,0,0) \tag{4.1}$$

where the position $(x,y,\bar{z})$ indicates the observation point while $(0, 0, 0)$ indicates the origin of the wave. A method to find the true position of the object $\bar{z}$ is outlined later. Then, the received scattered wave $E^{\text{sc}}(x',y')$ due to the object when the transmitting and the receiving antennas are at $(x',y',0)$ and (x',y',D), respectively, can be expressed as

$$E^{\text{sc}}(x',y',D) = \int\int_{x\,y} f(x,y,\bar{z}) \cdot \frac{E_0^{\text{inc}}(x-x',y-y',\bar{z}) \cdot e^{-jk\sqrt{(x'-x)^2 + (y'-y)^2 + (D-\bar{z})^2}}}{\sqrt{(x'-x)^2 + (y'-y)^2 + (D-\bar{z})^2}} dxdy \tag{4.2}$$

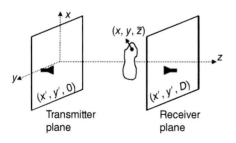

Figure 4.1 2D near-field microwave holography setup [128].

where k is the wave number in the background medium and $f(x,y,\bar{z})$ is the reflectivity function of the object. We denote with $g_0(x,y,\bar{z})$ the point-scattering function of (4.2), i.e.

$$g_0(x,y,\bar{z}) = \frac{E_0^{inc}(-x,-y,\bar{z}) \cdot e^{-jk\sqrt{x^2+y^2+(D-\bar{z})^2}}}{\sqrt{x^2+y^2+(D-\bar{z})^2}} \qquad (4.3)$$

Note that $g_0(x',y',\bar{z})$ is exactly the system PSF, i.e. the response $E^{sc}(x',y',D)$ to a point scatterer (the reflectivity $f(x,y,\bar{z})$ is a δ-function) at the center of the imaged plane $x = 0$, $y = 0$, $z = \bar{z}$. Next, E^{sc} is expressed as a 2D convolution integral:

$$E^{sc}(x',y',D) = \iint_{x\,y} f(x,y,\bar{z}) \cdot g_0(x'-x,y'-y,\bar{z})dxdy. \qquad (4.4)$$

This allows for its expression in 2D Fourier space as

$$\tilde{E}^{sc}(k_x,k_y,D) = \tilde{F}(k_x,k_y,\bar{z}) \cdot \tilde{G}_0(k_x,k_y,\bar{z}) \qquad (4.5)$$

where $\tilde{F}(k_x,k_y,\bar{z})$ and $\tilde{G}_0(k_x,k_y,\bar{z})$ are the 2D FT of $f(x,y,\bar{z})$ and $g_0(x,y,\bar{z})$, respectively. The Fourier variables k_x and k_y correspond to x and y, respectively. Finally, the reconstructed reflectivity function of the object is obtained as

$$f(x,y,\bar{z}) = F_{2D}^{-1}\left\{ \frac{\tilde{E}^{sc}(k_x,k_y)}{\tilde{G}_0(k_x,k_y,\bar{z})} \right\}. \qquad (4.6)$$

The reconstructed image can then be obtained as the magnitude of the reflectivity function $|f(x,y,\bar{z})|$.

Note that (4.6) is a formal reconstruction formula and it can be seen as the "maximum likelihood solution" if the collected data suffer from incompleteness or any particular type of noise. The latter factors are often the cause for ill-posedness. However, as shown in [127], the technique is very robust to noise for the imaging applications considered here, which is mainly due to sufficient sampling (in the Nyquist sense) along the scanned apertures. Also, because of the finite size of the apertures, not all the wave numbers (k_x,k_y) can be measured. As discussed later, this imposes lower and upper limits on k_x and k_y, which in turn limit the cross-range resolution of the images.

4.1.1 Using All Reflection and Transmission S-Parameters

In the imaging setup shown in Figure 4.1, four S-parameters can be measured for the two antennas at each scanning position: (i) s_{11} is a measure of the scattered wave received by antenna 1 when the same antenna is transmitting, (ii) s_{21} is a measure of the scattered wave received by antenna 2 when antenna 1 is transmitting, (iii) s_{12} is a measure of the scattered wave received by antenna 1 when antenna 2 is transmitting, and (iv) s_{22} is a measure of the scattered wave received by antenna 2 when the same antenna is transmitting. The parameters s_{11} and s_{22} are referred to as the reflection S-parameters for antenna 1 and antenna 2, respectively, whereas s_{21} and s_{12} are the transmission S-parameters. The 2D FT of s_{11}, s_{21}, s_{12}, and s_{22} are denoted as $\tilde{S}_{11}$, $\tilde{S}_{21}$, $\tilde{S}_{12}$, and $\tilde{S}_{22}$, respectively.

The collected data from each S-parameter can be processed using (4.6) to create an image of the object. In this case, we obtain four separate images. Note that in a passive two-port network such as the one consisting of two passive antennas and the object between them, we have $s_{12} = s_{21}$ due to reciprocity, even if the two antennas are different. In such a case, we have only three independent sets of S-parameters. Of course, such scenario would lead to two identical images obtained as a result of the image processing on $\tilde{S}_{21}$ and $\tilde{S}_{12}$. However, if the measured two-port network contains amplifiers or other nonlinear and/or non-reciprocal devices, we do have four independent data sets.

To reconstruct a single image from all four (or three) S-parameter data sets, simultaneous processing is needed. To implement this, we rewrite (4.5) for each S-parameter at each (k_x,k_y), assuming that $\tilde{F}(k_x,k_y,\bar{z})$ is the same in all cases. This leads to

$$
\begin{bmatrix}
\tilde{S}_{11}(k_x,k_y) \\
\tilde{S}_{21}(k_x,k_y) \\
\tilde{S}_{12}(k_x,k_y) \\
\tilde{S}_{22}(k_x,k_y)
\end{bmatrix}
= \tilde{F}(k_x,k_y,\bar{z})
\begin{bmatrix}
\tilde{G}_0^{11}(k_x,k_y,\bar{z}) \\
\tilde{G}_0^{21}(k_x,k_y,\bar{z}) \\
\tilde{G}_0^{12}(k_x,k_y,\bar{z}) \\
\tilde{G}_0^{22}(k_x,k_y,\bar{z})
\end{bmatrix}.
\tag{4.7}
$$

This system of equations is over-determined since at each (k_x,k_y), four equations need to be solved simultaneously to find only one unknown, $\tilde{F}(k_x,k_y,\bar{z})$. Thus, a least-square solution is employed at each spatial frequency pair (k_x,k_y) to find the corresponding value of $\tilde{F}(k_x,k_y,\bar{z})$. After $\tilde{F}(k_x,k_y,\bar{z})$ is obtained for all values of (k_x,k_y), $f(x,y,\bar{z})$ is computed as the inverse 2D FT of $\tilde{F}(k_x,k_y,\bar{z})$.

4.1.2 Localization of the Object Along the Range

As discussed in the previous section, the true position of the object $\bar{z}$ along the range (the z axis) is a prerequisite for obtaining the scattering function $g_0(x,y,\bar{z})$ and consequently its 2D FT $\tilde{G}_0(k_x,k_y,\bar{z})$, which is needed in the reconstruction formula (4.6). Therefore, a method is needed to determine $\bar{z}$. The method discussed here is based on the images created from the reflection S-parameters of the two antennas.

The 2D FT of the reflectivity functions at an arbitrary selected position z reconstructed from reflection S-parameters using (4.6), $\tilde{F}^1\left(k_x,k_y,z\right)$ and $\tilde{F}^2\left(k_x,k_y,z\right)$, can be written as

$$\tilde{F}^1\left(k_x,k_y,z\right) = \frac{\tilde{S}_{11}\left(k_x,k_y,\bar{z}\right)}{\tilde{G}_0^1\left(k_x,k_y,z\right)} \tag{4.8}$$

$$\tilde{F}^2\left(k_x,k_y,z\right) = \frac{\tilde{S}_{22}\left(k_x,k_y,\bar{z}\right)}{\tilde{G}_0^2\left(k_x,k_y,z\right)} \tag{4.9}$$

where $z = \bar{z}$ emphasizes that the 2D FT of the measured reflection parameters are associated with the true position of the object $\bar{z}$. $\tilde{G}_0^1\left(k_x,k_y,z\right)$ and $\tilde{G}_0^2\left(k_x,k_y,z\right)$ are the FTs of the scattering functions for antenna 1 and antenna 2, respectively. They are computed based on the assumed object position z when the antennas are at the origin of their respective aperture planes (as described by (4.3)).

A cost function $J(z)$ is formulated as the 2-norm of the difference between the 2D FT of the two reconstructed reflectivity functions:

$$J(z) = \left\| \tilde{F}^1\left(k_x,k_y,z\right) - \tilde{F}^2\left(k_x,k_y,z\right) \right\|. \tag{4.10}$$

We expect that the true position of the object would be the value of z for which $J(z)$ is minimized, i.e.

$$\bar{z} = \arg\min_z \left(J(z)\right). \tag{4.11}$$

In other words, the reflectivity functions obtained from the reflection coefficients of the two antennas are anticipated to be the same when the object position is estimated correctly.

To prove this assumption, we rewrite $J(z)$ using (4.8) and (4.9):

$$J(z) = \left\| \frac{\tilde{S}_{11}\left(k_x,k_y,\bar{z}\right)}{\tilde{G}_0^1\left(k_x,k_y,z\right)} - \frac{\tilde{S}_{22}\left(k_x,k_y,\bar{z}\right)}{\tilde{G}_0^2\left(k_x,k_y,z\right)} \right\|. \tag{4.12}$$

If we assume that the true (unknown) position of the object is $z = \bar{z}$, $\tilde{S}_{11}\left(k_x,k_y,\bar{z}\right)$ and $\tilde{S}_{22}\left(k_x,k_y,\bar{z}\right)$ can be expanded as

$$\tilde{S}_{11}\left(k_x,k_y,\bar{z}\right) = \tilde{F}\left(k_x,k_y,\bar{z}\right)\tilde{G}_0^1\left(k_x,k_y,\bar{z}\right) \tag{4.13}$$

$$\tilde{S}_{22}\left(k_x,k_y,\bar{z}\right) = \tilde{F}\left(k_x,k_y,\bar{z}\right)\tilde{G}_0^2\left(k_x,k_y,\bar{z}\right) \tag{4.14}$$

where $\tilde{F}\left(k_x,k_y,\bar{z}\right)$ is the true reflectivity function of the object, which is expected to be the same for both cases. Using (4.12)–(4.14), $J(z)$ is written as

$$J(z) = \left\| \tilde{F}\left(k_x, k_y, \bar{z}\right) \left[\frac{\tilde{G}_0^1\left(k_x, k_y, \bar{z}\right)}{\tilde{G}_0^1\left(k_x, k_y, z\right)} - \frac{\tilde{G}_0^2\left(k_x, k_y, \bar{z}\right)}{\tilde{G}_0^2\left(k_x, k_y, z\right)} \right] \right\|.$$ (4.15)

Assuming spherical propagation for the incident wave ($E_0^{\text{inc}}(x,y,z) = e^{-jkr}$ where r is the distance between the origin of the wave and the observation point), the scattering functions for the two antennas are

$$g_0^1(x,y,z) = \frac{e^{-j2kr_1}}{r_1}$$ (4.16)

$$g_0^2(x,y,z) = \frac{e^{-j2kr_2}}{r_2}$$ (4.17)

where $r_1 = \sqrt{x^2 + y^2 + z^2}$ and $r_2 = \sqrt{x^2 + y^2 + (D-z)^2}$. The 2D FT of (4.16) and (4.17) is [129]

$$\tilde{G}_0^1(k_x, k_y, z) = \frac{e^{-jk_z z}}{ik_z}$$ (4.18)

and

$$\tilde{G}_0^2(k_x, k_y, z) = \frac{e^{-jk_z(D-z)}}{ik_z}$$ (4.19)

where $k_z = \sqrt{4k^2 - k_x^2 - k_y^2}$. Using (4.18) and (4.19), the cost function in (4.15) is written as

$$J(z) = \left\| \tilde{F}\left(k_x, k_y, \bar{z}\right) \left[e^{-jk_z(\bar{z}-z)} - e^{-jk_z(z-\bar{z})} \right] \right\| = \left\| 2j\tilde{F}\left(k_x, k_y, \bar{z}\right) \sin\left[k_z(z-\bar{z}) \right] \right\|.$$ (4.20)

For the cost function of (4.20) to be minimized, the following must hold: $\sin\left[k_z(z-\bar{z})\right] = 0$ or $k_z(z-\bar{z}) = n\pi$ where $n \in \mathbb{Z}$. Therefore, $(z-\bar{z}) = n\lambda_z/2$, where $\lambda_z = 2\pi/k_z$ is the wavelength associated with k_z. Since k_z depends on k_x and k_y, λ_z also depends on k_x and k_y. This indicates that for each point $\left(k_x^0, k_y^0\right)$ in the 2D spatial-frequency domain, the minimum of the difference between $\tilde{F}^1\left(k_x^0, k_y^0, z\right)$ and $\tilde{F}^2\left(k_x^0, k_y^0, z\right)$ repeats every $\lambda_z/2$ when moving away from the true position of the object $\bar{z}$. Since $\lambda_z/2$ is not the same for different $\left(k_x^0, k_y^0\right)$ points, the minimum of the difference between $\tilde{F}^1\left(k_x, k_y, z\right)$ and $\tilde{F}^2\left(k_x, k_y, z\right)$ occurs only at $n = 0$ or $z = \bar{z}$, where all $\left(k_x^0, k_y^0\right)$ points give a common minimum.

The accuracy of the object localization along the range depends on the sampling rate in the range direction. As long as at least one of the sample planes falls inside the object volume, the described technique can identify the object's range location correctly by the minimum of the cost function. Note that in this single-frequency 2D reconstruction technique, the goal is not to build an image in depth but rather to identify planes, which contain the object. The 2D reconstruction assumes that the object is thin, i.e. its dimension along the range is much smaller than its dimensions in the cross-range directions. The obvious restriction here is that for proper reconstruction, the object must lie in a plane more or less parallel to the apertures scanned by the antennas.

4.1.3 Image Reconstruction Results

To demonstrate the performance of 2D near-field microwave holography technique, two $\lambda/2$ horizontally polarized dipole antennas at 35 GHz with an object in between are simulated. The arrangement is shown in Figure 4.2. The antennas perform a 2D scan by moving together in their respective aperture planes, apertures 1 and 2. The aperture planes are located at $z = 50$ mm and $z = 0$. The apertures have a size of 60 mm × 60 mm with their centers being on the z axis. The "X" shape object is parallel to the x–y plane. The S-parameters are calculated and recorded for every (x,y) position on apertures 1 and 2. The same setup is simulated without the object to obtain the background S-parameters. Then, the calibrated S-parameters at each (x,y) position are calculated by subtracting the background S-parameters from the S-parameters of the test simulation. Since the dipoles are horizontally polarized along the x axis, the x component of the simulated incident electric field E_x^{inc} is recorded at the imaged plane and is considered as $E_0^{inc}(x,y,\bar{z})$. We assume that the object has a relative permittivity of $\varepsilon_r = 5$ and conductivity of $\sigma = 0$ S/m. It is located in free space. Figure 4.3 shows the 2D

Figure 4.2 The X-shape object scanned by two horizontally-polarized dipoles at 35 GHz [128].

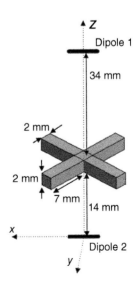

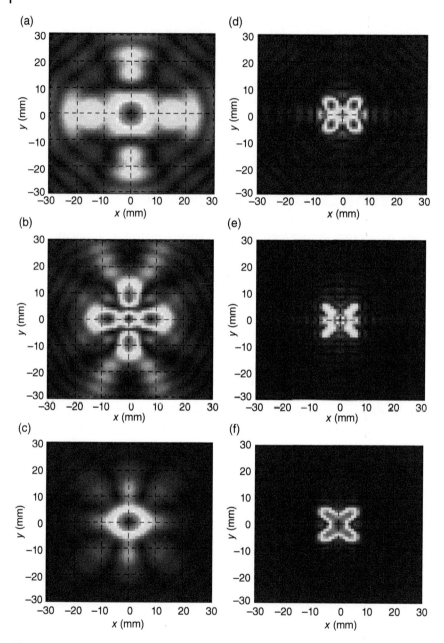

Figure 4.3 S-parameter plots in the imaging of a dielectric object ($\varepsilon_r = 5$ and $\sigma = 0$ S/m) in free space: (a) $|\tilde{S}_{11}^{cal}|$, (b) $|\tilde{S}_{22}^{cal}|$, (c) $|\tilde{S}_{12}^{cal}|$. Reconstructed images after applying the 2D holographic imaging technique (d) $|\tilde{S}_{11}^{cal}|$, (e) $|\tilde{S}_{22}^{cal}|$, (f) $|\tilde{S}_{12}^{cal}|$ [128].

color-map plots of the calibrated *S*-parameter magnitudes and the respective reconstructed images of the object. All reconstructed images reproduce the shape of the object accurately. In contrast, the 2D color maps of the *S*-parameter magnitudes do not provide any clue about the shape of the object.

4.2   3D Near-Field Holographic Imaging Using Incident Field and Green's Function

In this section, we discuss how the 2D near-field holographic imaging technique in [127] is extended to full 3D imaging [86] when wideband information is available. The method in [86] has a number of distinct features and advantages compared to the 3D holographic techniques in [5, 11]. First, the method allows for incorporating forward-scattered signals in addition to the backscattered signals. This additional information leads to more accurate reconstruction results and also allows for the significant suppression of image artifacts in the range direction. Second, the method allows for an incident-field distribution represented in numeric form. This distribution can be obtained either through simulation or measurement with the particular antenna setup and medium. Third, it also allows for numeric input of the Green function, i.e. the set of signals due to point scatterers in the given medium and received by the given antennas. These can be efficiently obtained through simulation as explained later. The accurate representations of the incident field and the Green function for the particular problem at hand are crucial in near-field imaging where analytical approximations such as plane or spherical waves are not adequate. Fourth, the numerical formats of the incident field and Green's function necessitate a new inversion procedure. The 3D holography methods discussed in Chapter 3 rely on the analytical (exponential) form of the incident field and Green's function in order to cast the inversion expression in the form of a 3D inverse FT. Re-sampling of the data in k_z-space is also necessary, which may lead to errors. This procedure is inapplicable with numeric representations of the incident field and Green's function. Instead, in [86] a system of equations is solved in each spatial frequency pair (k_x, k_y) and then 2D inverse FT is applied to the least-square solution at each desired range location. The systems of equations have much smaller dimensions and better condition number compared to the systems of equations in regular optimization-based microwave imaging techniques. The 3D object is reconstructed as a set of 2D slice images in parallel planes along range. Fifth, the algorithm is fast, accurate, and robust to high levels of noise. The scalar near-field holographic 3D imaging in [86] has been extended to a vectorial form in [130]. We describe this technique in detail next.

Consider a setup which consists of two antennas performing a planar scan and an object in-between as shown in Figure 4.4. With proper calibration, the

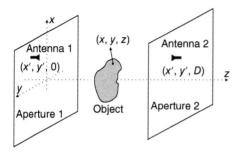

Figure 4.4 3D near-field microwave holography setup [86].

acquired signals represent samples of the field scattered by the object. As discussed previously, the forward model of scattering used by microwave holography makes use of the linear Born approximation [86], the integral form of which is given by

$$E^{sc}(r_P) \approx \iiint\limits_{V_Q} [k_s^2(r_Q) - k_b^2]\underline{\underline{G}}(r_P, r_Q) \cdot E^{inc}(r_Q) dr_Q \tag{4.21}$$

where E^{sc} is the scattered field, $\underline{\underline{G}}$ is Green's dyadic function, E^{inc} is the incident field, k_s and k_b are the wave numbers of the scatterer and the background medium, respectively, and V_Q is the inspected volume. We assume that k_b is known. The position vectors r_P and r_Q give the locations of the observation and scattering points, respectively.

As shown in Figure 4.4, the antennas perform a 2D scan while moving together (aligned along each other's boresight) on two separate parallel planes positioned at $z = 0$ and $z = D$.

Assume that at every measurement frequency ω_l ($l = 1, 2, ..., N_\omega$) we know the incident field $E^{inc}(0, 0, 0; x, y, z; \omega_l)$ at any point $r_Q = (x, y, z)$ in the inspected volume when the transmitting antenna is at $(0,0,0)$. In addition, assume that all components of Green's dyadic $G_i^j(x,y,z;0,0,D;\omega_l)$, $i, j = x, y, z$, are known for an i-polarized point source at (x, y, z) and a j-polarized response E_j^{sc} at $(0, 0, D)$. For brevity, we introduce the notations:

$$E_0^{inc}(x,y,z,\omega_l) \equiv E^{inc}(0,0,0;x,y,z;\omega_l) \tag{4.22}$$

$$\underline{\underline{G}}_0(x,y,z,\omega_l) \equiv \underline{\underline{G}}(x,y,z;0,0,D;\omega_l). \tag{4.23}$$

Let $E_j^{sc}(x',y',\omega_l)$, $j = x, y, z$, be the j-component of the scattered E-field to which the receiver at r_P responds, where $r_P = (x', y', D)$ for the forward-scattered wave and $r_P = (x', y', 0)$ for the backscattered wave. This implies that the transmitting antenna is at $(x', y', 0)$ since it moves together with the receiving antenna.

In a homogeneous or layered medium, where the layers are in x–y planes, the incident field and the Green tensor for the case where the antenna pair is at (x', y') can be obtained from (4.22) and (4.23) by a simple translation:

$$E^{inc}(x',y',0;x,y,z;\omega_l) = E_0^{inc}(x-x',y-y',z,\omega_l) \tag{4.24}$$

$$\underline{\underline{G}}(x',y',0;x,y,z;\omega_l) = \underline{\underline{G}}_0(x-x',y-y',z,\omega_l). \tag{4.25}$$

It follows from (4.21) that each component of the scattered field at the frequency ω_l is written as

$$E_j^{sc}(x',y',\omega_l) \approx \int\int_{z\ y\ x}\int f(x,y,z,\omega_l) \cdot \sum_{i=x,y,z} a_i^j(x-x',y-y',z,\omega_l)dxdydz \tag{4.26}$$

where

$$f(x,y,z,\omega_l) = k_s^2(x,y,z,\omega_l) - k_b^2(z,\omega_l) \tag{4.27}$$

$$a_i^j(x,y,z,\omega_l) = E_i^{inc}(x,y,z,\omega_l)G_{0i}^j(x,y,z,\omega_l). \tag{4.28}$$

Notice that k_b is independent of x and y but may depend on z. The contrast function $f(x, y, z, \omega_l)$ is the object of reconstruction. Here, the integral over x and y can be interpreted as a 2D convolution integral. Thus, the 2D FT of both side of (4.26) gives

$$\tilde{E}_j^{sc}(k_x,k_y,\omega_l) = \int_z \tilde{F}(k_x,k_y,z,\omega_l) \sum_{i=x,y,z} \tilde{A}_i^j(k_x,k_y,z,\omega_l)dz \tag{4.29}$$

where $\tilde{F}(k_x,k_y,z,\omega_l)$ and $\tilde{A}_i^j(k_x,k_y,z,\omega_l)$ are the 2D FTs of $f(x, y, z, \omega_l)$ and $a_i^j(x,y,z,\omega_l)$, respectively; k_x and k_y are the Fourier variables corresponding to x and y, respectively.

To reconstruct the contrast function, we first approximate the integral in (4.29) by a discrete sum with respect to z for N_z reconstruction planes:

$$\tilde{E}_j^{sc}(k_x,k_y,\omega_l) = \sum_{n=1}^{N_z} \tilde{F}(k_x,k_y,z_n,\omega_l) \sum_{i=x,y,z} \tilde{A}_i^j(k_x,k_y,z_n,\omega_l)\Delta z \tag{4.30}$$

where Δz is the distance between two neighboring reconstruction planes.

Since we perform measurements at N_ω frequencies, writing (4.30) for all frequencies leads to N_ω equations at each spatial frequency pair $\kappa = (k_x,k_y)$ as

$$\left\{ \begin{aligned} \tilde{E}_j^{sc}(\kappa,\omega_1) &\approx \sum_{i=x,y,z} \tilde{A}_i^j(\kappa,z_1,\omega_1)\tilde{F}(\kappa,z_1,\omega_1)\Delta z + \cdots + \sum_{i=x,y,z} \tilde{A}_i^j(\kappa,z_{N_z},\omega_1)\tilde{F}(\kappa,z_{N_z},\omega_1)\Delta z \\ &\qquad\qquad\qquad\vdots \\ \tilde{E}_j^{sc}(\kappa,\omega_{N_\omega}) &\approx \sum_{i=x,y,z} \tilde{A}_i^j(\kappa,z_1,\omega_{N_\omega})\tilde{F}(\kappa,z_1,\omega_{N_\omega})\Delta z + \cdots + \sum_{i=x,y,z} \tilde{A}_i^j(\kappa,z_{N_z},\omega_{N_\omega})\tilde{F}(\kappa,z_{N_z},\omega_{N_\omega})\Delta z. \end{aligned} \right. \tag{4.31}$$

In (4.31), we have N_ω decoupled equations from which finding directly the frequency-dependent function $\tilde{F}(\kappa, z_n, \omega_l)$, $n = 1, 2, \ldots, N_z$ and $l = 1, 2, \ldots, N_\omega$, is not feasible. However, under certain assumptions, a system of equations can be constructed from which the contrast function is found. First, we assume that the contrast function can be expressed as

$$f(x,y,z,\omega) = h(x,y,z)\phi(\omega) \tag{4.32}$$

where the variation of the contrast function with space and frequency is separated into the functions $h(x, y, z)$ and $\phi(\omega)$, respectively. Second, we assume that the variation of the contrast function with frequency $\phi(\omega)$ is known *a priori*. This function can be approximated for example by fitting Cole-Cole or Debye models to the dielectric properties of the background medium. Then, $\phi(\omega)$ would approximate the frequency behavior of both the background medium and the objects. In many applications where the operating frequency band is narrow (or in a nondispersive mediums), we assume $\phi(\omega) = 1$. Taking the 2D FT of both sides of (4.32) with respect to x and y leads to

$$\tilde{F}(\kappa, z, \omega) = \tilde{H}(\kappa, z)\phi(\omega) \tag{4.33}$$

where $\tilde{H}(\kappa, z)$ is the 2D FT of $h(x, y, z)$. Thus, replacing $\tilde{F}(\kappa, z, \omega)$ with $\tilde{H}(\kappa, z)\phi(\omega)$ in (4.31) leads to the following system of equations:

$$\begin{cases} y_1 = a_{11}x_1 + \cdots + a_{1N_z}x_{N_z} \\ \quad\vdots \\ y_{N_\omega} = a_{N_\omega 1}x_1 + \cdots + a_{N_\omega N_z}x_{N_z} \end{cases} \tag{4.34}$$

where

$$y_l = \tilde{E}_j^{sc}(\kappa, \omega_l) \tag{4.35}$$

$$a_{ln} = \phi(\omega_l) \sum_{i=x,y,z} \tilde{A}_i^j(\kappa, z_n, \omega_l) \tag{4.36}$$

$$x_n = \tilde{H}(\kappa, z_n). \tag{4.37}$$

This system of equations is solved in the least-square sense to find $\tilde{H}(\kappa, z_n)$, $n = 1, 2, \ldots, N_z$, at each spatial frequency pair $\kappa = (k_x, k_y)$. Tikhonov regularization [131] can be employed to reduce the ill-conditioning of the least-square solution in (4.34).

After solving the systems of equations at all points (k_x, k_y) in Fourier space, inverse 2D FT is applied to $\tilde{H}(\kappa, z_n)$, $n = 1, 2, \ldots, N_z$, to reconstruct a 2D slice of the function $h(x, y, z_n)$ at each $z = z_n$ plane. Then, the normalized modulus of $h(x, y, z_n)$, $|h(x, y, z_n)|/M$, where M is the maximum of $|h(x, y, z_n)|$ for all z_n, is plotted versus the spatial coordinates x and y to obtain a 2D image of the object at each $z = z_n$ plane. By putting together all 2D slice images, a 3D image of the object is obtained.

In the reconstruction process proposed in [86], it is assumed that at each frequency ω_l ($l = 1, 2, \dots, N_\omega$), the incident field $E_0^{inc}(x,y,z,\omega_l)$ and the Green's function $\underline{G}_0(x,y,z,\omega_l)$ are known at any object point (x, y, z) when the transmitting and the receiving antennas are at $(0, 0, 0)$ and $(0, 0, D)$, respectively. In Chapter 3, these had been assumed to be scalar functions approximated as $E^{inc}(x, y, z, \omega_l) = G_0(x, y, z, \omega_l) \approx e^{-jkr}$, which is only suitable for far-field scattering problems. In near-field imaging, these functions are obtained via simulations of the setup when scattering objects are absent. For example, to obtain $E_x^{inc}(x,y,z,\omega)$, the transmitting antenna is placed at the origin of the corresponding aperture plane and the x-component of the electric field is recorded in all reconstruction planes and at all frequencies of interest, i.e. $E_x^{inc}(x,y,z_n,\omega_l)$, $n = 1, 2, \dots, N_z$ and $l = 1, 2, \dots, N_\omega$.

One approach to obtaining the elements of Green's dyadic function, for example $G_{0x}^x(x,y,z,\omega_l)$, is to move an x-polarized point source on all reconstruction planes along all pixels and to record the field at the terminals of the receiving antenna at the center of the acquisition plane. This approach, however, is extremely inefficient because it requires as many simulations as the number of imaged voxels. Another more efficient approach is using the reciprocity principle [109]. Only one simulation needs to be carried out, wherein each receiving antennas is placed at the center of the acquisition plane and it operates in a transmitting mode. The electric field it generates is then sampled simultaneously at all pixels of all reconstruction planes.

The collected data set from each S-parameter can be processed separately to create an image of the object. In this case, we can obtain four separate images at each $z = z_n$ plane. Another approach is to reconstruct a single image from the data sets of all S-parameters. To implement this, we write (4.34) to get N_ω equations using the data collected from each S-parameter at each (k_x, k_y) point in Fourier space. Bearing in mind that the unknown distribution $\tilde{H}(\kappa, z_n)$ (or x_n in (4.34)), $n = 1, 2, \dots, N_z$, is the same in all four systems of equations, these are combined to construct a single system of equations with $4 \times N_\omega$ rows, every N_ω rows corresponding to one of the S-parameters. This system of equations is usually over-determined since at each (k_x, k_y), $4 \times N_\omega$ equations need to be solved simultaneously to find N_z unknowns $\tilde{H}(\kappa, z_n)$, $n = 1, 2, \dots, N_z$, where usually $N_z < N_\omega$. Again, a least-square solution is sought to find these unknowns.

4.2.1 Image Reconstruction Results

To demonstrate the accuracy of this 3D microwave holography technique, a setup with two $\lambda/2$ (at the center frequency of operation) x-polarized dipole antennas with objects in-between has been simulated in [86]; see Figure 4.5.

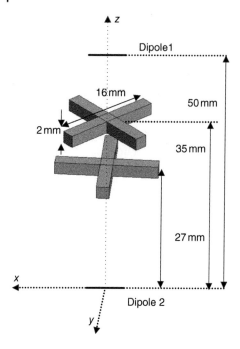

Figure 4.5 Dielectric objects in free space scanned by two $\lambda/2$ (at 35 GHz) horizontally polarized (x-polarized) dipoles; dipole 1 is moving on the $z = 50$ mm plane while dipole 2 is moving on the $z = 0$ mm plane. The simulated S-parameters are recorded in the frequency band of 25–45 GHz for two similar X-shape objects with square cross-sections 2 mm on a side and length of each arm 16 mm, parallel to the x–y plane, one centered at (0,0,27) mm with its arms along the x and y axes, the other one centered at (0,0,35) mm, with the arms rotated by 45° with respect to the x and y axes, both objects having dielectric properties of $\varepsilon_r = 2$ and $\sigma = 0$ S/m [86].

The antennas perform a 2D scan by moving together on two parallel aperture planes and collecting wideband data at each position. The apertures have a size of 60 mm × 60 mm with their centers being on the z axis.

The total S-parameters s^t_{jk} $(j,k = 1,2)$, i.e. the S-parameters in the presence of the scattering object, are acquired on planar surfaces with the dimensions mentioned above and with sampling steps of 1 mm along both x and y directions. Since the dipoles are x-polarized, the x-component of the simulated incident electric field E^{inc}_x and the Green's function G^x_x are recorded at the reconstruction planes $z = z_n$. These planes are of size 40 mm × 40 mm.

To evaluate the quality of the shape reconstruction at each $z = z_n$ plane, a reconstruction error (RE) is defined as

$$\text{RE}(z_n) = \frac{1}{N_x N_y} \sum_{i=1}^{N_x} \sum_{j=1}^{N_y} \left\| |f(x_i, y_j, z_n)| / M - \bar{I}(x_i, y_j, z_n) \right\| \tag{4.38}$$

where $\| \cdot \|$ is the L_1 norm, N_x and N_y are the number of samples along the x and y directions, respectively, $|f(x_i, y_j, z_n)|$ is a slice of the image at z_n and M is the maximum contrast value found in the entire set of slices $|f(x_i, y_j, z_n)|$, $n = 1, 2, \ldots, N_z$. The function $I(x_i, y_j, z_n)$ represents the true magnitude of the object's contrast normalized to its maximum value. It is between 0 and 1 inside the object (depending on the contrast of the object) and 0 elsewhere. Also, to have an overall estimate of the RE, we define the parameter RE^t as

$$\mathrm{RE}^{t} = \frac{1}{N_z} \sum_{n=1}^{N_z} \mathrm{RE}(z_n). \tag{4.39}$$

In Figure 4.5, an "X" shape object parallel to the x–y plane is placed at $z = 27$ mm. The "X" shape has two orthogonal arms of length 16 mm and square cross-sections 2 mm on a side. Another similar "X" shape object is rotated by 45° about the z axis and is placed at $z = 35$ mm. The constitutive parameters of the cuboids are $\varepsilon_r = 2$ and $\sigma = 0$ S/m whereas the background is free space. The distance between the two objects is 8 mm, which is very close to the range resolution limit $\delta_z = 7.5$ mm. Figure 4.6 shows the reconstructed images. The objects are recovered well at $z = 27$ mm and $z = 35$ mm. The images formed at other range locations show weak artifacts. Later, we also discuss an approach to reduce these artifacts. For the object at $z = 27$ mm, the arms along the x axis have an apparent reflectivity, which is stronger than that of the arms along the y axis. This is due to the fact that the dipoles are oriented along x and thus the incident field is x-polarized. Such field interacts better with the arms oriented along x.

It is worthwhile to compare this technique to the holography technique in [5] (discussed in Section 3.3.2). Figure 4.7 shows the reconstructed images for the objects in Figure 4.5 when applying the 3D holography technique presented in [5]. Note that this technique makes use of the reflection (backscattered) signals

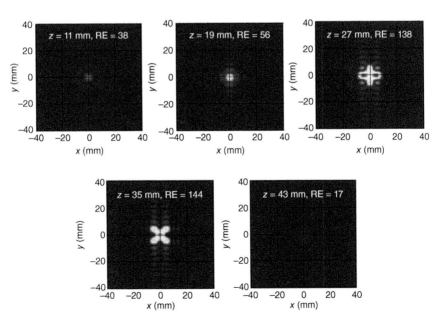

Figure 4.6 Reconstructed images when using all S-parameters simultaneously for the dielectric objects in Figure 4.5 ($\mathrm{RE}^{t} = 393$) [86].

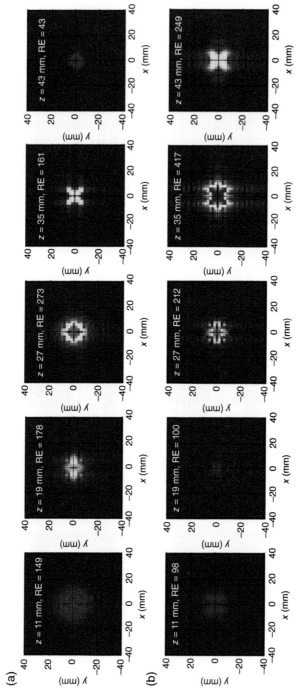

Figure 4.7 Reconstructed images obtained for the dielectric objects in Figure 4.5 when applying the 3D holography technique in [5] to: (a) s_{11}^{cal} ($RE^t = 804$) and (b) s_{22}^{cal} ($RE^t = 1076$) [86].

only. The same number of frequencies is adopted and the reconstructed images are shown on the same planes. As observed, artifacts are present and the object shapes are not recovered well. For the sake of comparison, Figure 4.8 shows the reconstructed images obtained from near-field holographic imaging technique discussed in this section where only the reflection S-parameters are processed. The comparison of the RE and RE^t values for the images in Figure 4.7 with those in Figure 4.8 confirms the advantages of using adequate representation of the incident field and Green's function in near-field imaging instead of using the plane-wave assumptions as in [5]. In addition, the possibility of using transmission S-parameters in conjunction with the reflection ones in the near-field imaging technique leads to even better results as shown in Figure 4.6. The values of RE and RE^t confirm this observation.

In order to investigate the effect of random noise on the near-field 3D holography technique, we consider two types of noise: (i) white Gaussian noise added to the complex values of the S-parameters (such noise could be produced by electronic devices or the environment) and (ii) white Gaussian noise added only to the phase of the S-parameters (such noise could be due to the mechanical vibrations of the antennas).

Figure 4.9 shows the reconstructed images for the objects in Figure 4.5 when adding noise of type 1 and 2, respectively. For SNR value as low as -10 dB for the first type of noise and 10 dB for the second type, the quality of the reconstructed images is still comparable to that of the images obtained from noiseless data (see Figure 4.6) although RE and RE^t degrade. This indicates that the reconstruction technique is robust to noise. SNR values lower than -10 dB and 10 dB for noise of type 1 and 2, respectively, lead to visible degradation of the reconstructed images.

4.2.2 Suppressing Artifacts Along Range

To improve the quality of the image reconstruction, the object localization concept discussed in Section 4.1.2 can be extended to 3D imaging for removing the artifacts in images. The process involves the following steps. First, using only the s_{11} data, the images $|h_{11}(x, y, z_n)|$, $n = 1, 2, \ldots, N_z$, are obtained. Second, using only the s_{22} data, the images $|h_{22}(x, y, z_n)|$, $n = 1, 2, \ldots, N_z$, are obtained. Third, the $|h_{11}(x, y, z_n)|$ and $|h_{22}(x, y, z_n)|$ images are subtracted at each reconstruction plane and the 2-norm of the difference is taken as

$$J(z_n) = \left\| \frac{|h_{11}(x,y,z_n)|}{\max|h_{11}(x,y,z_n)|} - \frac{|h_{22}(x,y,z_n)|}{\max|h_{22}(x,y,z_n)|} \right\|_2, n = 1, 2, \ldots, N_z. \quad (4.40)$$

Assuming that the artifacts are not similar in $|h_{11}(x, y, z_n)|$ and $|h_{22}(x, y, z_n)|$ unlike the objects, which should appear consistently in both images, we expect that the function in (4.40) has minima in the true positions of the objects. This can be employed to suppress the artifacts as explained next.

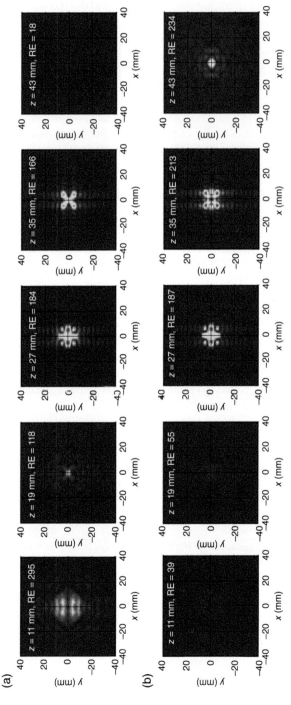

Figure 4.8 Reconstructed images obtained for the dielectric objects in Figure 4.5 when applying the 3D holography technique to: (a) s_{11}^{cal} ($RE^t = 781$) and (b) s_{22}^{cal} ($RE^t = 728$) [86].

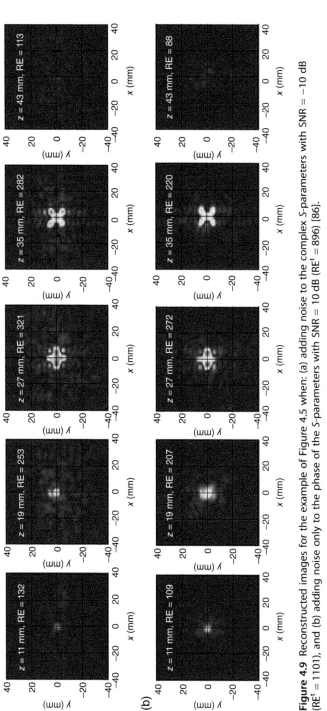

Figure 4.9 Reconstructed images for the example of Figure 4.5 when: (a) adding noise to the complex *S*-parameters with SNR = −10 dB (RE^t = 1101), and (b) adding noise only to the phase of the *S*-parameters with SNR = 10 dB (RE^t = 896) [86].

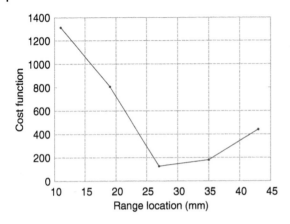

Figure 4.10 The variation of cost function in (4.40) for the example of Figure 4.6 [86].

The function $J(z)$ can be used to weigh properly the slice images along the range. Images at range positions where $J(z)$ is small are assigned larger weight than those where $J(z)$ is large. This is implemented using

$$| \hat{h}(x,y,z_n) | = \frac{|h(x,y,z_n)|}{J(z_n)}, n = 1,2,...,N_z \qquad (4.41)$$

where $| \hat{h}(x,y,z_n) |$ is the enhanced image.

As an example, Figure 4.10 shows the cost function (4.40) computed for the example of Figure 4.5. It has the lowest values at $z = 27$ mm and $z = 35$ mm, where the actual objects are. From the reconstructed images shown in Figure 4.6, an object is also observed at $z = 19$ mm. However, the large value of the cost function at this range location indicates that this object is an artifact. Figure 4.11 shows the enhanced images $| \hat{h}(x,y,z_n) |$. Note that the RE and RE^{t} values reduce and the artifact seen at $z = 19$ mm in Figure 4.6 disappears.

4.3 Microwave Holographic Imaging Employing Forward-Scattered Waves Only

The first experimental near-field holographic imaging results were presented in [109]. There, the main focus has been on the application of breast-cancer diagnostics and imaging of artificial glycerin-based phantoms emulating the breast tissues. Only 2D holographic imaging results were presented employing forward-scattered waves. This is because the backscattered (or reflected) signals are very weak in tissue measurements and they are masked by the noise and uncertainties. On the other hand, the backscattered signals are crucial for implementing 3D holographic imaging with planar scanning. The images

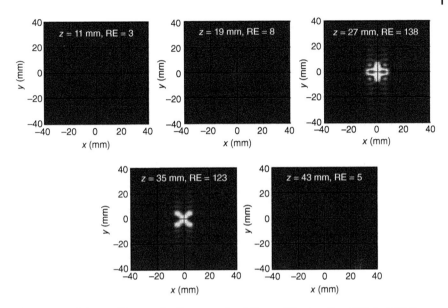

Figure 4.11 Enhanced images for the example of Figure 4.5 using the artifact removal method ($RE^t = 277$) [86].

obtained when the backscattered data are excluded do not have depth resolution. Moreover, if 3D reconstruction is attempted without backscattered signals, the image quality in the cross range is compromised, although a 2D holographic reconstruction in a slice of the object is successful. Overall, without the backscattered data, the holographic images contain strong artifacts along both range and cross range. In [110], a configuration has been proposed which allows for 3D holographic imaging where only forward-scattered data are available. The configuration in [110] is a scanning setup comprising one transmitter and five receivers which move together during the 2D scan on two rectangular apertures on opposite sides of the inspected region. It was shown that 3D image reconstruction is possible with this setup when employing only forward-scattered data. Besides, in [110], range and cross-range resolution limits have been derived using an approach developed for bistatic SAR [132].

4.3.1 Resolution in a Two-Antenna Configuration

In [110], first, a two-antenna microwave holography setup is considered where the antennas scan simultaneously two rectangular planar surfaces in a raster pattern. This setup is illustrated in Figure 4.12 where antenna 1 and antenna 2 perform the scan together on aperture 1 at $z = 0$ and aperture 2 at $z = \bar{z}$. At each sampling step, the wideband transmission coefficient of this two-port

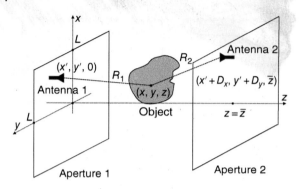

Figure 4.12 Microwave holography setup for through measurements [110].

system is acquired, which represents the forward-scattered wave. The backscattered waves are ignored since in many applications in lossy media, e.g. tissue imaging, these waves are too weak and are likely to be masked by noise and uncertainties in the measurements. Also, the measurement system may be nonreciprocal due to the use of amplifiers (or other nonreciprocal components) at the transmitting and/or receiving sides. This makes the measurement of the reflected signals impossible unless separate Tx and Rx channels are provided.

In [86, 109], both antennas are assumed to be always in the same (x', y') position during the scan. In [110], a general case has been considered where antenna 2 may have an offset of $\pm D_x$ and $\pm D_y$ with respect to antenna 1 along the x- and y-axes, respectively.

In the following, the spatial resolution limits of the two-antenna imaging setup is studied employing an approach proposed for bistatic SAR [132]. Note that the resolution limits derived with this method (which assumes propagating waves) provide good estimates in the case of the two-antenna holographic setup where the object is in the far-field region of the antennas. However, in the near-field imaging it is possible to obtain better resolution due to the availability of evanescent waves. The closer the antennas are to the object, the better the resolution is in a near-field imaging scenario.

With reference to Figure 4.12, at each sampling position $(x', y', 0)$ for antenna 1 – this implies that antenna 2 is at $(x' + D_x, y' + D_y, \bar{z})$ – the total path traveled by the wave through the object is denoted by R and is written as

$$R = R_1 + R_2 \tag{4.42}$$

where

$$R_1 = \sqrt{(x - x')^2 + (y - y')^2 + z^2} \tag{4.43}$$

$$R_2 = \sqrt{[(x' + D_x) - x]^2 + [(y' + D_y) - y]^2 + (\bar{z} - z)^2}. \tag{4.44}$$

The differential change in R, denoted by ΔR, due to an incremental change $(\Delta \rho_s \hat{s})$ in the position of the object in any arbitrary direction $\hat{s}$, is written as [132]

$$\Delta R = (\hat{s}^T \cdot \nabla R) \Delta \rho_s \tag{4.45}$$

where ∇ is the gradient operator, and the superscript T represents the transpose operator. Here, $\hat{s}$ is a unit vector.

On the other hand, the minimum "measureable" ΔR is related to the velocity of the wave in the medium c and the bandwidth of the imaging system B as [132]

$$\Delta R_{min} \approx \frac{c}{B}. \tag{4.46}$$

From (4.45) and (4.46), the resolution limit in the direction $\hat{s}$, denoted by $\bar{\rho}_s$, is estimated as [132]

$$\bar{\rho}_s \approx \frac{c}{B} \cdot \frac{1}{|\hat{s}^T \nabla R|}. \tag{4.47}$$

Note that the resolution here is defined as the half-power (or half-intensity) width of the image of a point-like object.

To estimate the resolution of the imaging system along the x, y, and z directions, first the gradient of R in a rectangular coordinate system is written as

$$\nabla R = \begin{bmatrix} \dfrac{\partial R}{\partial x} \\[2mm] \dfrac{\partial R}{\partial y} \\[2mm] \dfrac{\partial R}{\partial z} \end{bmatrix} = \begin{bmatrix} \dfrac{(x-x')}{R_1} - \dfrac{[(x'+D_x)-x]}{R_2} \\[2mm] \dfrac{(y-y')}{R_1} - \dfrac{[(y'+D_y)-y]}{R_2} \\[2mm] \dfrac{z}{R_1} - \dfrac{(\bar{z}-z)}{R_2} \end{bmatrix}. \tag{4.48}$$

Then, from (4.47) and (4.48), the cross-range resolution $\bar{\rho}_u$ ($u = x$ or y) and the range resolution $\bar{\rho}_z$ are obtained as

$$\bar{\rho}_u = \frac{c}{B} \cdot \frac{1}{\left| \dfrac{(u-u')}{R_1} - \dfrac{[(u'+D_u)-u]}{R_2} \right|}, u = x, y \text{ and } u' = x', y' \tag{4.49}$$

$$\bar{\rho}_z = \frac{c}{B} \cdot \frac{1}{\left| \dfrac{z}{R_1} - \dfrac{(\bar{z}-z)}{R_2} \right|}. \tag{4.50}$$

In the above equations, $\bar{\rho}_u$ and $\bar{\rho}_z$ are obtained when $\hat{s}$ in (4.47) is substituted with the unit vector in the particular direction, x, y, or z.

We assume that the object is very small and it is positioned on the z-axis, i.e. the object is at $(0, 0, z)$. Then, (4.49) and (4.50) are simplified as

$$\bar{\rho}_u = \frac{c}{B} \cdot \frac{1}{\left| \frac{-u'}{R_1'} - \frac{(u' + D_u)}{R_2'} \right|}, u' = x', y' \text{ and } D_u = D_x, D_y \tag{4.51}$$

$$\bar{\rho}_z = \frac{c}{B} \cdot \frac{1}{\left| \frac{z}{R_1'} - \frac{(\bar{z} - z)}{R_2'} \right|} \tag{4.52}$$

where

$$R_1' = \sqrt{x'^2 + y'^2 + z^2} \tag{4.53}$$

$$R_2' = \sqrt{(x' + D_x)^2 + (y' + D_y)^2 + (\bar{z} - z)^2}. \tag{4.54}$$

We observe from the above equations that the range and cross-range resolution limits depend on the positions of the transceivers and the object.

To derive the resolution limits, a 2D cut in the 3D setup in Figure 4.12 is considered as shown in Figure 4.13, where the axis u represents either x or y. D_u and u' denote the offsets of the receiver with respect to the transmitter and the position of the transmitter, respectively.

With reference to Figure 4.13, we first rewrite the resolution limits in (4.51) and (4.52) in terms of the angles θ_1 and θ_2 as

$$\bar{\rho}_u = \frac{c}{B} \cdot \frac{1}{|\sin\theta_1 + \sin\theta_2|} \tag{4.55}$$

$$\bar{\rho}_z = \frac{c}{B} \cdot \frac{1}{|\cos\theta_1 - \cos\theta_2|}. \tag{4.56}$$

These are the angles of incidence and scattering, respectively, associated with the imaged point-like object.

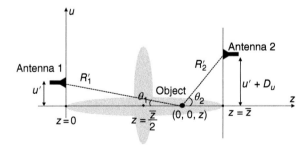

Figure 4.13 Simplifying the 3D setup in Figure 4.12 to a 2D view for deriving the resolution limits in terms of θ_1 and θ_2. When $D_u = 0$, the range resolution is poor in the shaded region [110].

From (4.56), it follows that when the transmitter and the receiver are at the same position (x', y') during the 2D scan (as in [86, 109]), i.e. when $D_u = 0$, there are two main regions in the inspected volume where the range resolution $\bar{\rho}_z$ is poor. These two regions are: (i) the middle of the inspected domain, i.e. $z \approx \bar{z}/2$, and (ii) the direct path connecting the transmitting and receiving antennas passing right through the scatterer, i.e. $\theta_1 \approx \theta_2 \approx 0$. In Figure 4.13, the regions of poor range resolution are shaded in gray.

On the other hand, when the receiver has an offset with respect to the transmitter, i.e. when $D_u \neq 0$, we obtain finite values for both the cross-range and the range resolution limits everywhere in the inspected volume. In this configuration, as per (4.55), the best cross-range resolution is obtained when the antennas are at the edges of the apertures, i.e. the angles θ_1 and θ_2 are close to $\pi/2$:

$$\bar{\rho}_u^{min}\Big|_{\theta_1 = \theta_2 = \pm \frac{\pi}{2}} = \frac{c}{2B}. \tag{4.57}$$

On the other hand, according to (4.55), the cross-range resolution is best when one antenna is on the z-axis ($\theta_1 = 0$) while the other antenna is at $u \to \infty$ ($D_u \to \infty, \theta_2 \to \pi/2$):

$$\bar{\rho}_u = \frac{c}{B}. \tag{4.58}$$

Under these same conditions, the range resolution is given by

$$\bar{\rho}_z^{min}\Big|_{\substack{\theta_1 = 0, \theta_2 = \pm \pi/2 \\ \theta_1 = \pm \pi/2, \theta_2 = 0}} = \frac{c}{B}. \tag{4.59}$$

It is worth noting that in reflection holography where $\theta_2 = \pi - \theta_1$, the cross-range and range resolution limits are

$$\bar{\rho}_u = \frac{c}{B} \cdot \frac{1}{|2\sin\theta_1|} \tag{4.60}$$

$$\bar{\rho}_z = \frac{c}{B} \cdot \frac{1}{|2\cos\theta_1|}. \tag{4.61}$$

Assuming $B \approx 2f_c$ with f_c being the center frequency of the band B, the cross-range resolution can be written as

$$\bar{\rho}_u = \frac{\lambda_c}{4\sin\theta_1} \tag{4.62}$$

where λ_c is the wavelength at f_c. Thus, in this case, the cross-range resolution improves if the size of the aperture is large ($\theta_1 \to \pi/2$). On the other hand, the best range resolution limit is obtained when $\theta_1 \to 0$ as

$$\bar{\rho}_z^{min} = \frac{c}{2B}. \tag{4.63}$$

These results are consistent with the expressions for the resolution limits of reflection holography in [5] (discussed in Section 3.3.2.2).

4.3.2 Multiple Receiver Setup

In the previous section, it was shown that having a nonzero offset distance for the receiver along the x or y axis ($D_u \neq 0$, $u = x, y$) leads to improving the range and cross-range resolutions in a scenario where only forward-scattered signals are acquired. Thus, in [110] a multiple receiver setup has been proposed in a star distribution to achieve satisfactory range and cross-range resolutions in 3D microwave imaging with planar raster scanning.

Figure 4.14 illustrates the setup where one transmitter illuminates the object while five receivers measure the forward-scattered waves. Antenna 2 is aligned with the transmitter (antenna 1) while the other receivers (antennas 3–6) have offset distances of $\pm D_x$ and $\pm D_y$ along the $\pm x$ and $\pm y$ directions, respectively. The transmitter and the five receivers move together during the 2D scan on the two planar apertures. From the results in the previous section, it follows that larger offset distances lead to improved resolution. However, at large offset distances the scattered wave travels along longer paths from the object to the receiver. This weakens the signal due to two factors: spatial spread and attenuation if the medium is lossy. Besides, signal strength can also weaken due to the antenna pattern. These factors impose upper limits on the offset distances. In addition, increasing the offset distances increases the size of aperture 2. If the size of aperture 1 is $2L_x \times 2L_y$, then the size of aperture 2 is $2(L_x + D_x) \times 2(L_y + D_y)$.

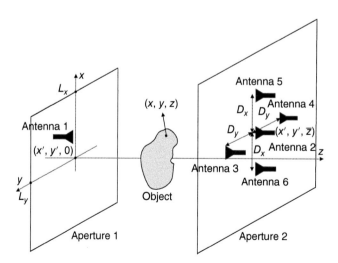

Figure 4.14 3D near-field microwave holography setup using forward-scattered waves only. Antenna 1 illuminates the object while antennas 2–6 receive the scattered waves [110].

4.3.3 Holographic Image Reconstruction

In this section, the 3D holographic microwave imaging algorithm in [86] is extended to process the data obtained with the multiple receiver configuration. With reference to Figure 4.14, the transmitting antenna (antenna 1) and the receiving antenna array (antennas 2–6) perform a 2D scan at the planes $z = 0$ and $z = \bar{z}$, respectively. For simplicity, we assume that the antennas are x-polarized (e.g. dipoles oriented along the x axis). Thus, the field can be reasonably approximated by a TM_x polarization and the theory is scalar in nature. The approach here is directly applicable to acquisition of co- and cross-polarized data. In the scalar case, the Green's function can be viewed as the G_x^x element of the full dyadic while the E-field is represented by its x-component only. From now on we omit the subscript x for brevity.

Assume that at any measurement frequency ω_l ($l = 1, 2, \ldots, N_\omega$) the incident field $E^{inc}(0, 0, 0; x, y, z; \omega_l)$ is known at any point (x,y,z) in the inspected volume due to antenna 1 when it is at $(0,0,0)$. In addition, Green's function for antenna 2 $G_2(x,y,z;0,0,\bar{z};\omega_l)$ is known for an x-polarized scattering point source at (x, y, z) and the x-polarized response at $(0,0,\bar{z})$. This information can be obtained via simulations. For brevity, we introduce the notations

$$E_0^{inc}(x,y,z,\omega_l) \equiv E^{inc}(0,0,0;x,y,z;\omega_l) \tag{4.64}$$

$$G_2(x,y,z,\omega_l) \equiv G_2(x,y,z;0,0,\bar{z};\omega_l). \tag{4.65}$$

In [109], it has been shown that while E^{inc} can be obtained from simulations, Green's function can be obtained from E^{inc} using the reciprocity principle. Thus, assuming that antennas 1–6 are identical, only one simulation suffices to obtain $E^{inc}(x, y, z, \omega_l)$, from which Green's functions for antennas 2–6 are obtained as

$$G_2(x,y,z,\omega_l) = E^{inc}(x,y,\bar{z}-z,\omega_l) \tag{4.66}$$

$$G_3(x,y,z,\omega_l) = E^{inc}(x,y-D_y,\bar{z}-z,\omega_l) \tag{4.67}$$

$$G_4(x,y,z,\omega_l) = E^{inc}(x,y+D_y,\bar{z}-z,\omega_l) \tag{4.68}$$

$$G_5(x,y,z,\omega_l) = E^{inc}(x-D_x,y,\bar{z}-z,\omega_l) \tag{4.69}$$

$$G_6(x,y,z,\omega_l) = E^{inc}(x+D_x,y,\bar{z}-z,\omega_l). \tag{4.70}$$

Let $E_k^{sc}(x',y',\bar{z},\omega_l)$ be the scattered E-field received by the kth antenna ($k = 2, \ldots, 6$) when the transmitting antenna is at $(x', y', 0)$. Following the approach presented in Section 4.2, $E_k^{sc}(x',y',\bar{z},\omega_l)$ is written as

$$E_k^{sc}(x',y',\bar{z},\omega_l) = \iiint_{z\,y\,x} f(x,y,z) \cdot a_k(x'-x,y'-y,z,\omega_l)dxdydz, \text{ for } k = 2,\ldots,6 \tag{4.71}$$

where $f(x,y,z) = k_s^2(x,y,z) - k_b^2$ is the contrast function, k_s and k_b are the wave numbers of the object and the background media, respectively, and

$$a_k(x,y,z,\omega_l) = E^{\text{inc}}(-x,-y,z,\omega_l)G_k(-x,-y,z,\omega_l). \tag{4.72}$$

Notice that in (4.71), the integration over x and y can be interpreted as a 2D convolution integral. Thus, the 2D FT of E_k^{sc} is written as

$$\tilde{E}_k^{\text{sc}}(k_x,k_y,\bar{z},\omega_l) = \int_z \tilde{F}(k_x,k_y,z)\tilde{A}_k(k_x,k_y,z,\omega_l)dz \quad \text{for } k = 2,\dots,6 \tag{4.73}$$

where $\tilde{F}$ and $\tilde{A}_k$ are the 2D FTs of $f(x, y, z)$ and $a_k(x, y, z, \omega_l)$, respectively. By approximating the integral in (4.73) by a discrete sum and employing the data collected at all frequencies and at all receivers, we construct a system of equations at each $\boldsymbol{\kappa} = (k_x,k_y)$ as

$$\underline{\mathbf{E}} = \underline{\underline{\mathbf{A}}}\,\underline{\mathbf{F}} \tag{4.74}$$

where

$$\underline{\mathbf{E}} = \begin{bmatrix} \left[\underline{\tilde{E}}_2^{\text{sc}}\right] \\ \vdots \\ \left[\underline{\tilde{E}}_6^{\text{sc}}\right] \end{bmatrix}, \underline{\underline{\mathbf{A}}} = \begin{bmatrix} \left[\underline{\underline{\tilde{A}}}_2\right] \\ \vdots \\ \left[\underline{\underline{\tilde{A}}}_6\right] \end{bmatrix}, \underline{\tilde{\mathbf{F}}} = \begin{bmatrix} \tilde{F}(\boldsymbol{\kappa},z_1) \\ \vdots \\ \tilde{F}(\boldsymbol{\kappa},z_{N_z}) \end{bmatrix} \tag{4.75}$$

and for each $k = 2, \dots, 6$

$$\left[\underline{\tilde{E}}_k^{\text{sc}}\right] = \begin{bmatrix} \tilde{E}_k^{\text{sc}}(\boldsymbol{\kappa},\bar{z},\omega_1) \\ \vdots \\ \tilde{E}_k^{\text{sc}}(\boldsymbol{\kappa},\bar{z},\omega_{N_\omega}) \end{bmatrix} \tag{4.76}$$

$$\left[\underline{\underline{\tilde{A}}}_k\right] = \begin{bmatrix} \tilde{A}_k(\boldsymbol{\kappa},z_1,\omega_1)\Delta z & \dots & \tilde{A}_k(\boldsymbol{\kappa},z_{N_z},\omega_1)\Delta z \\ \vdots & \vdots & \vdots \\ \tilde{A}_k(\boldsymbol{\kappa},z_1,\omega_{N\omega})\Delta z & \dots & \tilde{A}_k(\boldsymbol{\kappa},z_{N_z},\omega_{N\omega})\Delta z \end{bmatrix}. \tag{4.77}$$

Here, Δz is the discretization step along the z-axis.

The system in (4.74) is solved at each (k_x,k_y) pair for $\tilde{F}(\boldsymbol{\kappa},z_n)$, $n = 1, \dots, N_z$, in a least-square sense. Once the systems of equations for all (k_x, k_y) are solved, the inverse 2D FT is applied to $\tilde{F}(\boldsymbol{\kappa},z_n)$, $n = 1, 2, \dots, N_z$, to reconstruct a 2D slice of the function $f(x, y, z_n)$ at each $z = z_n$ plane. Then, the normalized modulus of $f(x, y, z_n)$, $|f(x, y, z_n)|/M$, where M is the maximum of $|f(x, y, z_n)|$ for all z_n, is plotted versus the spatial coordinates x and y to obtain 2D images of the object at all N_z planes. By putting together all 2D slice images, a 3D image of the object is obtained.

4.4 Microwave Holographic Imaging Employing PSF of the Imaging System

In the near-field holographic imaging [86, 109, 110], the data for the incident field and Green's function are obtained via simulations. In practice, often, the fidelity of the simulation models, although better than the analytical approximations, may still be too low to ensure good image quality. The fidelity of the simulation model can be assessed through its ability to reproduce the measurements of known objects performed with the particular imaging setup. One factor contributing to the low fidelity of the simulations in near-field imaging can be the *numerical errors* such as those due to a coarse discretization mesh or imperfect absorbing boundary conditions. Such errors, however, are not the major concern because they can be reduced by mesh refinement and stricter convergence criteria for the numerical solution. Unfortunately, this may also lead to prohibitive computational burden. The major concern in near-field imaging is in the so called *modeling errors*, which are much more difficult to reduce. These errors are rooted in the inability to predict all influencing factors arising in the practical implementation of the acquisition setup. These include: fabrication tolerances of the antennas and the positioning components, uncertainties in the constitutive parameters of the materials (especially the absorbers) used to build the measurement chamber, deformations due to temperature or humidity, etc. In addition, models often ignore complexities in the cables, the connectors, the fine components of the measurement chamber (e.g. screws, brackets, thin supporting plates), etc.

In [87], a method of acquiring the incident field and Green's function specific to the particular acquisition system has been proposed via measurements of a known calibration object (CO). The method exploits the concept of PSF of a linear imaging system where the response due to an arbitrary object is the convolution of the response due to a point-like scatterer (the CO) with the spatial variation of that object. This technique is reviewed here together with some of the obtained results.

4.4.1 Using Measured PSF in Holographic Reconstruction

For a sufficiently small CO at the center $(0, 0, z_n)$ of the plane $z = z_n$, the contrast function in (4.27) can be written as (assuming nondispersive media)

$$f(x, y, z_n) \approx C\delta(x, y, z - z_n) \tag{4.78}$$

where C is the complex-valued CO permittivity contrast and $\delta(\cdot)$ is the Dirac delta function. As follows from (4.26), the field scattered by the CO and received by a j-polarized antenna is

$$E_j^{\text{sc,co}}(x', y', \omega_l; z_n) \approx \sum_{i=x,y,z} a_i^j(-x', -y', z_n, \omega_l). \tag{4.79}$$

The scattered field in (4.79) is obtained through measurements of the CO at all range distances (or layers) of interest z_n, $n = 1, \ldots, N_z$. It represents the PSF of the microwave measurement system at the respective range distances. From (4.26) and (4.79), the scattered field for any unknown large object can be expressed as

$$E_j^{sc}(x',y',\omega_l) \approx \sum_{n=1}^{N_z} \Delta z_n \iint\limits_{xy} f(x,y,z_n) \cdot E_j^{sc,co}(x'-x,y'-y,\omega_l;z_n)\,dy\,dx. \quad (4.80)$$

Here, the integral over z has been replaced by a discrete sum and Δz_n is the nth discretization interval in the range direction. Further, the integral over x and y can be interpreted as a 2D convolution. Thus, the 2D FT of $E_j^{sc}(x',y',\omega_l)$, $j = x, y, z$, is written as

$$\tilde{E}_j^{sc}(k_x,k_y,\omega_l) \approx \sum_{n=1}^{N_z} \Delta z_n \tilde{F}(k_x,k_y,z_n) \cdot \tilde{E}_j^{sc,co}(k_x,k_y,\omega_l;z_n) \quad (4.81)$$

where $\tilde{F}(k_x,k_y,z_n)$ and $\tilde{E}_j^{sc,co}(k_x,k_y,\omega_l;z_n)$ are the 2D FTs of $f(x, y, z_n)$ and $E_j^{sc,co}(x,y,\omega_l;z_n)$, respectively; and k_x and k_y are the respective Fourier variables.

To reconstruct the contrast function, (4.81) is applied to the data at all frequencies ω_l ($l = 1, \ldots, N_\omega$) to obtain a linear system of equations at each spatial frequency pair $\kappa = (k_x,k_y)$:

$$\begin{bmatrix} \tilde{E}_j^{sc}(\kappa,\omega_1) \\ \vdots \\ \tilde{E}_j^{sc}(\kappa,\omega_{N_\omega}) \end{bmatrix} \approx \begin{bmatrix} \tilde{E}_j^{sc,co}(\kappa,\omega_1;z_1)\Delta z_1 & \cdots & \tilde{E}_j^{sc,co}(\kappa,\omega_1;z_{N_z})\Delta z_{N_z} \\ \vdots & \vdots & \vdots \\ \tilde{E}_j^{sc,co}(\kappa,\omega_{N_\omega};z_1)\Delta z_1 & \cdots & \tilde{E}_j^{sc,co}(\kappa,\omega_{N_\omega};z_{N_z})\Delta z_{N_z} \end{bmatrix} \times \begin{bmatrix} \tilde{F}(\kappa,z_1) \\ \vdots \\ \tilde{F}(\kappa,z_{N_z}) \end{bmatrix}.$$
$$(4.82)$$

The system in (4.82), written for each κ, is solved in a least-square sense. Once the systems of equations for all κ are solved, the inverse 2D FT is applied to $\tilde{F}(\kappa,z_n)$, $n = 1, 2, \ldots, N_z$, to reconstruct a 2D slice of the contrast function $f(x, y, z_n)$ at each $z = z_n$ plane. Then, the normalized modulus of $f(x, y, z_n)$, $|f(x, y, z_n)|/M$, where M is the maximum of $|f(x, y, z_n)|$ for all z_n, is plotted versus x and y to obtain 2D images of the object at all N_z planes. By putting together all 2D slice images, a 3D image of the object is obtained. To this end, (4.82) is obtained with a single j-polarized receiving antenna aligned with the transmitting antenna where the two antennas scan together their respective planes.

4.4.2 Using Multiple Receivers in 3D Reconstruction

In a lossy background medium or in a nonreciprocal measurement setup, the backscattered waves may be very weak or not available. On the other hand, the availability of the backscattered waves is critical for the spatial resolution in the range direction with planar scanning where the transmitting and receiving antennas are aligned. The solution as discussed in the previous section is to use one transmitter antenna on one plane and multiple receiver antennas on the opposite plane that are shifted in the cross-range directions with respect to the transmitting antenna.

As discussed above, the product of the incident field with Green's function can be obtained from the measurements of a CO. Let $E_k^{sc}(x',y',\omega_l)$ and $E_k^{sc,co}(x',y',\omega_l;z)$ be the scattered E-field components received by the kth antenna ($k = 2, ..., N_k$), for the object under test (OUT) and the CO, respectively. Here, the subscript k indicates the receiver number and also implies the received field component since each antenna is characterized by its own polarization. Following the approach discussed above, a system of equations can be constructed at each spatial frequency pair $\boldsymbol{\kappa} = (k_x,k_y)$ that contains the data collected by all $N_k - 1$ receiving antennas:

$$\underline{\tilde{\mathbf{E}}}^{sc} = \underline{\underline{\tilde{\mathbf{D}}}}\,\underline{\tilde{\mathbf{F}}} \tag{4.83}$$

where

$$\underline{\tilde{\mathbf{E}}}^{sc} = \begin{bmatrix} \underline{\tilde{\mathbf{E}}}_2^{sc} \\ \vdots \\ \underline{\tilde{\mathbf{E}}}_{N_k}^{sc} \end{bmatrix}, \underline{\underline{\tilde{\mathbf{D}}}} = \begin{bmatrix} \underline{\underline{\tilde{\mathbf{D}}}}_2 \\ \vdots \\ \underline{\underline{\tilde{\mathbf{D}}}}_{N_k} \end{bmatrix}, \underline{\tilde{\mathbf{F}}} = \begin{bmatrix} \tilde{F}(\boldsymbol{\kappa},z_1) \\ \vdots \\ \tilde{F}(\boldsymbol{\kappa},z_{N_z}) \end{bmatrix} \tag{4.84}$$

$$\underline{\tilde{\mathbf{E}}}_k^{sc} = \begin{bmatrix} \tilde{E}_k^{sc}(\boldsymbol{\kappa},\omega_1) \\ \vdots \\ \tilde{E}_k^{sc}(\boldsymbol{\kappa},\omega_{N_\omega}) \end{bmatrix}, k = 2,...,N_k \tag{4.85}$$

$$\underline{\underline{\tilde{\mathbf{D}}}}_k = \begin{bmatrix} \tilde{E}_k^{sc,co}(\boldsymbol{\kappa},\omega_1;z_1)\Delta z_1 & \cdots & \tilde{E}_k^{sc,co}(\boldsymbol{\kappa},\omega_1;z_{N_z})\Delta z_{N_z} \\ \vdots & \vdots & \vdots \\ \tilde{E}_k^{sc,co}(\boldsymbol{\kappa},\omega_{N_\omega};z_1)\Delta z_1 & \cdots & \tilde{E}_k^{sc,co}(\boldsymbol{\kappa},\omega_{N_\omega};z_{N_z})\Delta z_{N_z} \end{bmatrix}. \tag{4.86}$$

The solution of the systems of equations at each $\boldsymbol{\kappa}$ and the formation of the images are similar to those discussed in the previous section.

4.4.3 Simulated Image Reconstruction Results

Figure 4.15 shows a simulation example. In this example, dielectric objects are immersed in a background medium with relative permittivity $\varepsilon_r = 16$ and conductivity $\sigma = 0.5$ S/m. They are scanned by two x-polarized dipoles, $\lambda/2$ long at

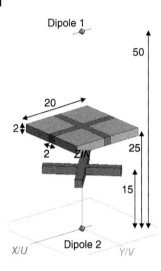

Dipole 1

Dipole 2

X/U   Y/V

Figure 4.15 Simulation example for near-field holographic imaging with PSF information. The background medium has relative permittivity $\varepsilon_r = 16$ and conductivity $\sigma = 0.5$ S/m whereas the X-shape objects are characterized by $\varepsilon_r = 32$ and $\sigma = 1.0$ S/m. The top X-shape object is embedded in a layer of $\varepsilon_r = 24$ and $\sigma = 0.75$ S/m. All dimensions are in mm [87].

6.5 GHz, where dipole 1 scans the $z = 50$ mm plane and dipole 2 scans the $z = 0$ mm plane. The simulated S-parameters are recorded in the frequency band between 3 and 11 GHz with a step of 0.25 GHz. The scattered field is acquired over two apertures of size 60 mm × 60 mm with x and y sampling steps of 1.5 mm, which are well below the limit of $\lambda_{min}/2$ discussed in Section 3.3.2.1. Here, $\lambda_{min} \approx 6.8$ mm. In both examples, the OUT consists of two identical X-shape objects with square cross-sections of 2 mm on a side and an arm length of 20 mm. The two objects are rotated with respect to each other by 45°. Both have dielectric properties of $\varepsilon_r = 32$ and $\sigma = 1$ S/m. Note that the X-shape object, positioned in the plane $z = 25$ mm, is embedded in a layer with material parameters $\varepsilon_r = 24$ and $\sigma = 0.75$ S/m. This layer is square, 20 mm on a side, and is 2 mm thick.

The cross-range and range resolution limits for these examples are found to be about 3 and 10 mm, respectively, using expressions discussed in Section 3.3.2.2. These expressions are strictly valid for far-zone imaging. Here, the dipoles are close to the scattering objects, i.e. at a distance comparable to the longest wavelength; therefore, the so found resolution limits are only approximate and, in fact, conservative. As the distance between the antennas and the object decreases, the angles supported by the object, as viewed by the two omnidirectional antennas, increase. This improves further the cross-range resolution [110]. Also, at such close ranges, the evanescent scattered field is expected to play a role in further improving the resolution. Taking into account the conservative range-resolution estimate of 10 mm, we choose a distance of 10 mm between the reconstruction planes along z.

A cubical perfect electric conductor (PEC) object with a side of 2 mm is used as the CO. First, the scattered fields due to this CO are obtained via full scans in five cases where the CO is placed at $(x,y) = (0,0)$ and at five range locations: $z = 5$,

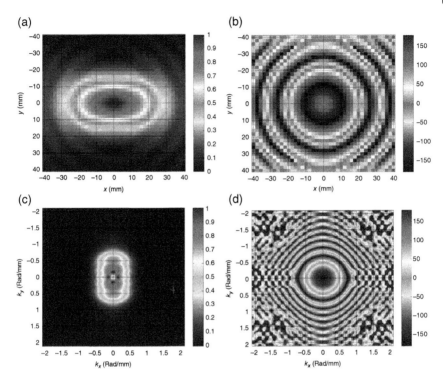

Figure 4.16 The magnitude and phase of the PSF for the examples in Figure 4.15 at 6.5 GHz and when the CO is at $z = 25$ mm: (a) magnitude of the PSF in the spatial domain, (b) phase of the PSF in the spatial domain, (c) magnitude of the PSF in the spectral domain, and (d) phase of the PSF in the spectral domain [87].

15, 25, 35, and 45 mm. Figure 4.16 shows the spatial and spectral variation of the collected PSF at 6.5 GHz when the CO is at $z = 25$ mm.

Then, the data are acquired with the two OUTs and the 3D holographic reconstruction is applied. Regularization is not employed because the simulated data are virtually noise-free and linearly independent. As a result, the matrices of the systems solved for all κ pairs, see (4.82), are well-conditioned. Figure 4.17 shows the reconstructed images and the corresponding RE and RE^t values. The two X-shape objects are reconstructed well in both examples. The low-contrast host medium is reconstructed as well.

4.4.4 Experimental Results with Open-Ended Waveguides

In this section, first, we show that using simulated incident field and Green's function as used in Section 3.3 may fail in the imaging of complicated objects due to the inaccuracies of the simulation models. For this purpose, we perform

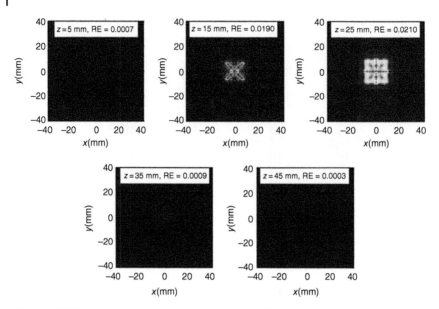

Figure 4.17 Reconstructed images obtained using a CO of size 2 mm for the dielectric objects in Figure 4.15; $RE^t = 0.0084$ [87].

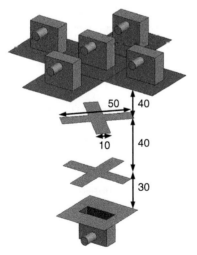

Figure 4.18 Illustration of the experimental imaging setup consisting of six X-band open-ended waveguides where one acts as a transmitter and five act as receivers. All dimensions are in mm. Two X-shaped metallic sheets are placed at $z = 30$ mm and $z = 70$ mm. The scan area is 160 mm by 200 mm and the sampling step along both x and y is 5 mm. The frequency range is from 3 to 20 GHz with a step of 0.25 GHz. The background medium is air [87].

imaging of two X-shape copper sheets with an arm length and width of 50 and 10 mm, respectively. The computer model of the imaging setup is shown in Figure 4.18 together with the two X-shape objects. Figure 4.19a shows the photo of the setup. One of the objects is positioned at $z = 30$ mm (the distance from the

(a) (b)

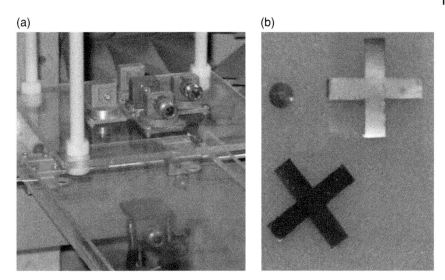

Figure 4.19 Measurement setup with the open-waveguide switched antenna array: (a) photo of the setup without scattering objects, and (b) photo showing the small scatterer used as a CO and the X-shape objects used to make the object under test (OUT) [87].

receiving antennas) with its arms aligned with the x and y axes whereas the second object is placed at $z = 70$ mm with its arms at $45°$ angle with respect to the x and y axes. The sampling in the spatial and frequency domains, the size of the scanned aperture, and the frequency range are summarized in the caption of Figure 4.18. The sampling step along x and y is 5 mm, which is below $\lambda_{min}/2$ ($\lambda_{min} = 15$ mm).

Figure 4.20 shows the reconstructed images with the corresponding RE and RE^t values when using the simulated incident field and Green's functions and without regularization. The quality of the images is not satisfactory and cannot be improved by the Tikhonov regularization.

Then, a CO, which is a small circular copper sheet with a diameter of 10 mm (shown in Figure 4.19b), is employed to measure the PSF at the range distances from $z = 10$ mm to $z = 100$ mm with a step of 10 mm. Figure 4.21 shows the reconstructed images and the corresponding RE and RE^t values when the approach is employed without regularization. It is observed that the quality of the reconstructed images is improved significantly compared to those in Figure 4.20. The X-shape objects are resolved well in both cross-range and range, and the image artifacts are relatively weak. If the Tikhonov regularization is employed, the images and the respective errors do not change significantly.

We note that the open-waveguide antennas are matched well to a 50-Ω impedance (reflection is below or about −6 dB) only in the band between

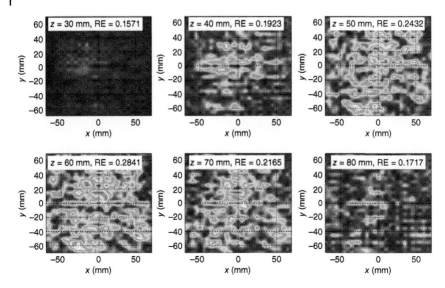

Figure 4.20 Reconstructed images of the two X-shape metallic objects (measured with the open-ended waveguide setup) obtained by using the simulated incident field distributions and Green's functions ($RE^t = 0.2108$). Regularization is not applied [87].

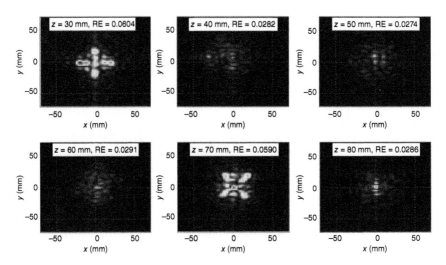

Figure 4.21 Reconstructed images of the two X-shape metallic objects (measured with the open-ended waveguide setup) obtained by using the measured PSF ($RE^t = 0.0388$). No regularization is applied [87].

approximately 8 and 18 GHz. The X-band waveguide section's cutoff frequency is 6.56 GHz. From 18 to 20 GHz, the antenna reflection coefficients are anywhere between −4 and −2 dB, while those from 3 to 6 GHz vary between −1 and 0 dB. Yet, the images in Figure 4.21 utilize the data acquired in the whole frequency band from 3 to 20 GHz. These images are very similar to the ones obtained using only the data above 8 GHz. However, if the low-frequency data are included, the artifacts along range are less pronounced, which is due to the increase in bandwidth. At the same time, the signal strength below 8 GHz is close to the noise floor of the system (more so below the waveguide cutoff). Thus, the approach, through the least-square solutions of (4.83) at each κ, effectively suppresses the "noisy" data sets and makes optimal use of the available data in the widest possible frequency band.

4.4.5 3D Imaging of Small Objects with the Bow-Tie Array

In this setup, the antenna in [133] is used as the transmitter and a bow-tie array sensor is used as the receiver. The scanned aperture has dimensions of 120 mm × 120 mm (the area scanned by the transmitter and the center element of the array).

The sampling steps along both x and y are 3 mm. They ensure sufficient sampling for a minimum wavelength in the background tissue phantom (with $\varepsilon_r = 10$ and $\tan \delta = 0.5$) of $\lambda_{\min} \approx 9.2$ mm. The frequency band of the measurement is from 3 to 10 GHz.

The thickness of the custom-made tissue phantom is 5 cm. To support the phantom and provide a flat surface for scanning, the tissue phantom is placed on a 5-mm thick printed circuit board (PCB) substrate, *Taconic CER 10* with $\varepsilon_r = 10$ and $\tan \delta = 0.0035$. There is also an air gap of almost 2 mm between the antennas and the phantom surfaces to prevent friction.

The dielectric properties of the tissue phantom used here as well as those of the substrate are practically frequency independent in the frequency band of the measurement.

Three objects are embedded inside the phantom. The objects are small cylindrical dielectric resonators with base diameter of 10 mm, height of 10 mm, and properties of $\varepsilon_r = 50$ and $\tan \delta \approx 0.001$. Two objects are placed in the $z = 15$ mm plane with center-to-center distance of approximately 15 mm and one object is in the center of the plane at $z = 35$ mm. Figure 4.22 shows a photo of the measurement setup.

An RF switch and a vector network analyzer (VNA) are employed to measure the wideband transmission S-parameters at each sampling position between the transmitter and the nine receiving antennas. The transmission S-parameters are recorded in the frequency band between 3 and 10 GHz with a step of 0.25 GHz.

The challenges in this example are: (i) the background phantom is very lossy ($\tan\delta \approx 0.5$) and the objects are small resulting in extremely weak signals; (ii) the

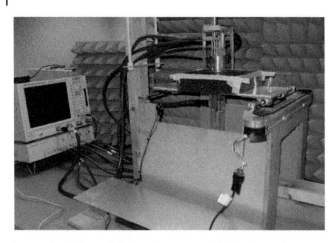

Figure 4.22 Photo of the imaging setup with the tissue phantom made of carbon-rubber sheets and the bow-tie switched sensor array (on top of the phantom). Illumination is provided by a TEM horn [133] (below the phantom) [87].

objects are positioned very close to each other so that the distance between them is approximately equal to the estimated resolution limit of holography; and (iii) the structure is three-dimensional.

First, Figure 4.24a shows the 3D holographic imaging results and the corresponding RE and RE^t values when the incident field and the Green's functions are obtained from simulations. The simulation of the incident field and Green's functions take extensive computational time and memory. For example, obtaining the incident field for each bow-tie element takes almost two days on an Intel(R) Xeon(R) CPU 2.93 GHz with 48 GB of RAM. Using such data, the objects are still not reconstructed well in the images due to the inaccuracies in the simulated field distributions. An indicator that such inaccuracies exist is the significant discrepancy between the simulated and measured reflection *S*-parameters of the bow-tie elements (shown in [87]). This discrepancy is mostly due to the modeling errors and is thus difficult to overcome by refining the simulation model.

In contrast, the simultaneous acquisition of the individual PSFs for all bow-ties in the array is done by simply performing the CO scans, where the CO is a small high-contrast dielectric cylinder (diameter is 10 mm, height is 10 mm, $\varepsilon_r = 50$, $\tan \delta \approx 0.001$). Figure 4.23 shows the CO embedded in the background medium.

Figure 4.24b shows the 3D holographic imaging results and the corresponding RE and RE^t values when the PSF calibration measurements are employed instead of simulations in order to obtain the required product of the incident field and the Green's function. The three dielectric objects are reconstructed

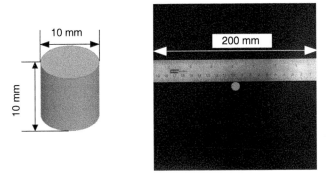

Figure 4.23 Diagram of the cylindrical dielectric resonator used as a calibration object (CO) in the experiments with tissue phantoms and a photo of the resonator embedded in a 200 mm x 200 mm layer of phantom material [87].

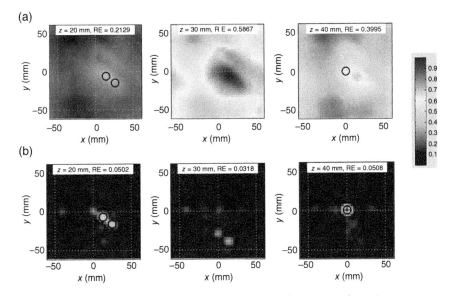

Figure 4.24 Reconstructed images in experiments using the receiving bow-tie array: (a) with the incident field and Green's function obtained via simulations ($RE^t = 0.3997$), and (b) with PSF calibration measurements ($RE^t = 0.0443$). The true position of the objects is shown with black circles. In both cases, Tikhonov's regularization is applied when solving (4.83) with a parameter value of 0.01 [87].

well. According to the spatial resolution limits derived in Section 4.3.1, the approximate range resolution is 14 mm. Since the range distance between the image slices is less than the range resolution, a weak shadow of each object is also observed in the slices adjacent to the planes in which the objects reside.

The comparison of these images with those in Figure 4.24a confirms that the fidelity of the measured incident-field and Green-function distributions is better than those obtained via simulations.

We note that in both sets of images (with simulated and measured PSFs), Tikhonov regularization with similar parameter value has been employed to solve the least-square problem in (4.83). Then, low-pass Gaussian filtering is applied to the resulting spectrum $\tilde{F}\left(k_x, k_y, z_n\right)$, $n = 1, 2, \ldots, N_z$, of the contrast function before performing the inverse FT. The low-pass filtering is beneficial when the data has low SNR, which is the case here. It suppresses the noise and the artifacts in the final images. However, its cutoff spatial frequency must be chosen so that the desired cross-range resolution is not compromised. Here, the spatial resolution δ is determined by the size of the small scattering object in the CO (10 mm). The cutoff spatial frequency of the Gaussian filter is then set to $k_c = \pi / \delta$, as per the relationship between the span in Fourier space and the cross-range resolution of holography [110] (At this cutoff point, the Gaussian filter should have very small level, for example, 5% of the peak value or lower).

4.4.6 Imaging of Large Objects with the Bow-Tie Array

Here, we demonstrate the imaging of the two X-shaped objects shown in Figure 4.19b, this time embedded in a 3-cm-thick lossy phantom with the same properties as in the previous section's example. One of the objects is placed at $z = 25$ mm with its arms aligned along the x and y axes, and the second objects is placed at $z = 35$ mm with its arms at 45° with respect to the x and y axes. The PSF was measured using the same 1 cm diameter copper circle shown in Figure 4.19b.

Figure 4.25a shows the 3D holographic imaging results and the corresponding RE and RE^t values when the incident field and Green's function are obtained through simulations. Figure 4.25b shows the images resulting from the reconstruction with the measured PSF. In both cases, no regularization is used as the resulting system matrices are well-conditioned. The low-pass filtering, however, is beneficial and is performed with the same cutoff as in the previous section's example. The two X-shaped objects are both reconstructed well when using the measured PSF. As in the previous section's example, this is not the case when the simulated field data are used.

4.5 3D Near-Field Holographic Imaging with Data Acquired over Cylindrical Apertures

Figure 4.26 illustrates a microwave imaging setup including an antenna that scans the backscattered field over a cylindrical aperture with radius of r_A and height of z_A. The backscattered waves are collected at N_ϕ number of angles evenly distributed within the interval $[0, 2\pi]$ and N_z number of positions evenly

(a)

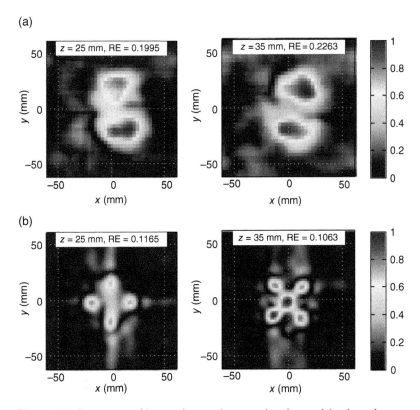

(b)

Figure 4.25 Reconstructed images in experiments using the receiving bow-tie array: (a) with the incident field and Green's function obtained via simulations ($RE^t = 0.2129$), and (b) with PSF calibration measurements ($RE^t = 0.1114$) [87].

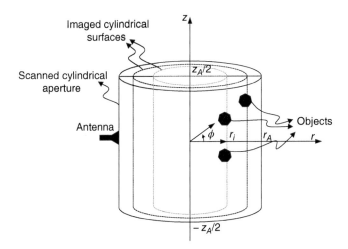

Figure 4.26 Illustration of a cylindrical microwave imaging setup. An antenna scans a cylindrical aperture and the images are reconstructed over cylindrical surfaces with radii $r = r_i$.

distributed within the interval $[-z_A/2,+z_A/2]$. At each sampling position (ϕ,z), the complex-valued scattered wave due to the object $E^{sc}(\phi, z)$ is measured at N_ω frequencies within the band ω_1 to ω_{N_ω}. The aim is to reconstruct images over cylindrical surfaces with radii r_i where $i = 1, ..., N_r$ and r_i is within $(0,r_A)$. For this purpose, the imaging system is assumed to be linear and space-invariant. This allows for using the approach discussed in Section 4.4 but in a cylindrical scenario. This approach relies on the measured PSF of the system to implement 3D imaging.

In order to collect the responses corresponding to the PSFs, first the responses $E^{sc,co}$ due to the COs placed at $(r_i,0,0)$, $i = 1, ..., N_r$, are recorded. These COs are placed one at a time at each $(r_i, 0, 0)$ position, $i = 1, ..., N_r$, while the scattered response due to each CO $E_i^{sc,co}(\phi,z)$ is measured over the cylindrical aperture by the antenna. Thus, the response due to any OUT at the cylindrical surface $r = r_i$, $E_i^{sc}(\phi,z)$, can be obtained using the convolution of the collected PSF for the corresponding surface with the contrast function for the object over that surface $f_i(\phi, z)$. This is written as

$$E_i^{sc}(\phi,z) = E_i^{sc,co}(\phi,z)*_\phi*_z f_i(\phi,z) \qquad (4.87)$$

where $*_\phi$ and $*_z$ denote convolutions with respect to the azimuthal (ϕ) and longitudinal (z) directions, respectively. Equation (4.87) is obtained following the same steps discussed for the rectangular setup in (4.78)–(4.80) but assuming spatial parameters ϕ and z instead of x and y. Besides, Eq. (4.87) can be written for the response created by an OUT over each imaged surface $(r_i,0,0)$, $i = 1, ..., N_r$. Then, the total response $E^{sc}(\phi, z)$ measured by the antenna due to the presence of objects on all the imaged surfaces can be approximated as the superposition of the contribution of all these responses $E_i^{sc}(\phi,z)$, $i = 1, ..., N_r$:

$$E^{sc}(\phi,z) = \sum_{i=1}^{N_r} E_i^{sc}(\phi,z) = \sum_{i=1}^{N_r} E_i^{sc,co}(\phi,z)*_\phi*_z f_i(\phi,z). \qquad (4.88)$$

In the above equation, the functions $E_i^{sc,co}(\phi,z)$ are known due to the measurement or simulation of the responses due to the COs and $E^{sc}(\phi,z)$ is known due to recording of the response of the objects. In order to estimate the unknown contrast functions $f_i(\phi,z)$, we perform measurements at multiple frequencies ω_l ($l = 1, ..., N_\omega$). Thus, (4.88) can be re-written at all these frequencies to provide the following system of equations:

$$\begin{cases} E^{sc}(\phi,z,\omega_1) = \displaystyle\sum_{i=1}^{N_r} E_i^{sc,co}(\phi,z,\omega_1)*_\phi*_z f_i(\phi,z) \\ \qquad\qquad\vdots \\ E^{sc}(\phi,z,\omega_{N_\omega}) = \displaystyle\sum_{i=1}^{N_r} E_i^{sc,co}(\phi,z,\omega_{N_\omega})*_\phi*_z f_i(\phi,z). \end{cases} \qquad (4.89)$$

In order to solve this system of equations we transform the equation to the spatial-frequency domain. In Section 4.4, doing this transformation was straightforward due to a planar aperture. However, here, the functions $E^{sc}(\phi, z, \omega_l)$, $E_i^{sc,co}(\phi, z, \omega_l)$, and $f_i(\phi, z)$ are periodic along the azimuthal direction (with a period of 2π) but aperiodic along the longitudinal direction. This requires special treatment.

In order to process the cylindrical data, we have to bear in mind that they are available in a discrete form as the spatially sampled versions of the scattered field. This indicates that the variations of the functions $E^{sc}(\phi, z, \omega_l)$, $E_i^{sc,co}(\phi, z, \omega_l)$, and $f_i(\phi, z)$ are represented by sequences of data samples $\mathbf{E}^{sc}(n_\phi, n_z, \omega_l)$, $\mathbf{E}_i^{sc,co}(n_\phi, n_z, \omega_l)$, and $\mathbf{f}_i(n_\phi, n_z)$, $n_\phi = 1, \dots, N_\phi$ and $n_z = 1, \dots, N_z$, taken at spatial and angular intervals denoted by Δz and $\Delta\phi$, respectively. Thus, the convolutions in (4.88) are written as

$$
\mathbf{E}^{sc}\left(n_\phi, n_z, \omega_l\right)
$$
$$
= \sum_{i=1}^{N_r} \mathrm{DTFT}_{z,\phi}^{-1}\left\{\mathrm{DTFT}_{z,\phi}\left\{\mathbf{E}_i^{sc,co}\left(n_\phi, n_z, \omega_l\right)\right\}\cdot\mathrm{DTFT}_{z,\phi}\left\{\mathbf{f}_i\left(n_\phi, n_z\right)\right\}\right\}
$$

$$(4.90)$$

where $\mathrm{DTFT}_{z,\phi}$ and $\mathrm{DTFT}_{z,\phi}^{-1}$ denote discrete-time Fourier transform (DTFT) and inverse DTFT along azimuthal and longitudinal directions, respectively. Taking DTFT from both sides of (4.90) leads to the following equation:

$$
\mathrm{DTFT}_{z,\phi}\left\{\mathbf{E}^{sc}\left(n_\phi, n_z, \omega_l\right)\right\}
$$
$$
= \sum_{i=1}^{N_r} \mathrm{DTFT}_{z,\phi}\left\{\mathbf{E}_i^{sc,co}\left(n_\phi, n_z, \omega_l\right)\right\}\cdot\mathrm{DTFT}_{z,\phi}\left\{\mathbf{f}_i\left(n_\phi, n_z\right)\right\}.
$$

$$(4.91)$$

The sequences $\mathbf{E}^{sc}(n_\phi, n_z, \omega_l)$, $\mathbf{E}_{r_i}^{sc,co}\left(n_\phi, n_z, \omega_l\right)$, and $\mathbf{f}_i(n_\phi, n_z)$ are aperiodic along z and we assume that N_z is sufficiently large such that the values of these sequences outside the observed window are negligible. Their DTFT is, however, a periodic function in the spatial frequency domain for the variable k_z which is the Fourier variable corresponding to z. The period of this DTFT function with respect to k_z is $1/\Delta z$. Furthermore, this periodic function is a summation of the FTs of the corresponding continuous functions $E_i^{sc,co}(\phi, z, \omega_l)$ and $f_i(\phi, z)$. If Δz is sufficiently small, the DTFTs of these functions within the range $[-1/(2\Delta z), 1/(2\Delta z)]$ can be observed and considered as the values of the continuous FTs of the functions $E_i^{sc,co}(\phi, z, \omega_l)$ and $f_i(\phi, z)$ with negligible distortion (aliasing) from the other terms.

The DTFT along the z variable for these sequences is written as

$$
\tilde{\mathbf{E}}^{sc}\left(n_\phi, k_z, \omega_l\right) \equiv \mathrm{DTFT}_z\left\{\mathbf{E}^{sc}\left(n_\phi, n_z, \omega_l\right)\right\} \approx \sum_{n_z=1}^{N_z} \mathbf{E}^{sc}\left(n_\phi, n_z, \omega_l\right)\exp\left(-jk_z n_z\right)
$$

$$(4.92)$$

$$\tilde{\mathbf{E}}_i^{\text{sc,co}}\left(n_\phi, k_z, \omega_l\right) \equiv \text{DTFT}_z\left\{\mathbf{E}_i^{\text{sc,co}}\left(n_\phi, n_z, \omega_l\right)\right\}$$

$$\approx \sum_{n_z=1}^{N_z} \mathbf{E}_i^{\text{sc,co}}\left(n_\phi, n_z, \omega_l\right) \exp\left(-jk_z n_z\right) \tag{4.93}$$

$$\tilde{\mathbf{f}}_i\left(n_\phi, k_z\right) \equiv \text{DTFT}_z\left\{\mathbf{f}_i\left(n_\phi, n_z\right)\right\} \approx \sum_{n_z=1}^{N_z} \mathbf{f}_i\left(n_\phi, n_z\right) \cdot \exp\left(-jk_z n_z\right) \tag{4.94}$$

where k_z is the Fourier variable with respect to z whereas $\tilde{\mathbf{E}}^{\text{sc}}\left(n_\phi, k_z, \omega_l\right)$, $\tilde{\mathbf{E}}_i^{\text{sc,co}}\left(n_\phi, k_z, \omega_l\right)$, and $\tilde{\mathbf{f}}_i\left(n_\phi, k_z\right)$ denote the DTFT along z (DTFT$_z$) for the sequences $\mathbf{E}^{\text{sc}}(n_\phi, n_z, \omega_l)$, $\mathbf{E}_i^{\text{sc,co}}\left(n_\phi, n_z, \omega_l\right)$, and $\mathbf{f}_i(n_\phi, n_z)$, respectively. On the other hand, since $\mathbf{E}_i^{\text{sc,co}}\left(n_\phi, n_z, \omega_l\right)$ and $\mathbf{f}_i(n_\phi, n_z)$ are periodic sequences along the azimuthal direction ϕ with a period of 2π, the convolution along the azimuthal direction ϕ is the circular convolution of these two sequences. The DTFT for the N_ϕ-periodic sequences along the azimuthal direction is computationally reduced to discrete Fourier transform (DFT) of these sequences, because: (i) Due to the periodicity of the functions in the spatial domain, in the spectral domain, the DTFT functions have samples only at frequencies equal to multiple integers of $1/(N_\phi\Delta\phi) = 1/(2\pi)$. These frequencies are known as harmonic frequencies. (ii) Due to the sampling step of $\Delta\phi$ in the spatial domain, the DTFT is periodic with a period of $1/\Delta\phi$. From the above-mentioned two properties, it is deduced that the maximum number of unique harmonic amplitudes is N_ϕ. Thus, DTFT in fact is reduced to DFT. This transformation with respect to the ϕ variable for the sequences $\tilde{\mathbf{E}}^{\text{sc}}\left(n_\phi, k_z, \omega_l\right)$, $\tilde{\mathbf{E}}_i^{\text{sc,co}}\left(n_\phi, k_z, \omega_l\right)$, and $\tilde{\mathbf{f}}_i\left(n_\phi, k_z\right)$ can be written as:

$$\tilde{\tilde{\mathbf{E}}}^{\text{sc}}\left(k_\phi, k_z, \omega_l\right) = \sum_{n_\phi=1}^{N_\phi} \tilde{\mathbf{E}}^{\text{sc}}\left(n_\phi, k_z, \omega_l\right) \cdot \exp\left[-j2\pi k_\phi\left(n_\phi-1\right)/N_\phi\right] \tag{4.95}$$

$$\tilde{\tilde{\mathbf{E}}}^{\text{sc,co}}\left(k_\phi, k_z, \omega_l\right) = \sum_{n_\phi=1}^{N_\phi} \tilde{\mathbf{E}}_i^{\text{sc,co}}\left(n_\phi, k_z, \omega_l\right) \cdot \exp\left[-j2\pi k_\phi\left(n_\phi-1\right)/N_\phi\right] \tag{4.96}$$

$$\tilde{\tilde{\mathbf{f}}}_i\left(k_\phi, k_z\right) = \sum_{n_\phi=1}^{N_\phi} \tilde{\mathbf{f}}_i\left(n_\phi, k_z\right) \cdot \exp\left[-j2\pi k_\phi\left(n_\phi-1\right)/N_\phi\right] \tag{4.97}$$

where k_ϕ is an integer varying from 0 to $N_\phi - 1$ with a step of 1. Using the transformations in (4.92)–(4.97) at all the frequencies leads to the following system of equations at each (k_ϕ, k_z) pair:

$$
\left\{
\begin{array}{l}
\tilde{\tilde{\mathbf{E}}}^{\text{sc}}\left(k_\phi, k_z, \omega_1\right) = \sum_{i=1}^{N_r} \tilde{\tilde{\mathbf{E}}}_i^{\text{sc,co}}\left(k_\phi, k_z, \omega_1\right) \tilde{\tilde{\mathbf{f}}}_i\left(k_\phi, k_z\right) \\
\qquad\qquad \vdots \\
\tilde{\tilde{\mathbf{E}}}^{\text{sc}}\left(k_\phi, k_z, \omega_{N_\omega}\right) = \sum_{i=1}^{N_r} \tilde{\tilde{\mathbf{E}}}_i^{\text{sc,co}}\left(k_\phi, k_z, \omega_{N_\omega}\right) \tilde{\tilde{\mathbf{f}}}_i\left(k_\phi, k_z\right).
\end{array}
\right.
\tag{4.98}
$$

Such systems of equations can be solved at each (k_ϕ, k_z) pair to obtain the values for $\tilde{\tilde{\mathbf{f}}}_i\left(k_\phi, k_z\right)$, $i = 1, ..., N_r$. Then, inverse DTFT along z and inverse DFT along ϕ can be applied to reconstruct the images of $\mathbf{f}_i(n_\phi, n_z)$ on all cylindrical surfaces $r = r_i$, $i = 1, ..., N_r$.

4.5.1 Imaging Results

This imaging technique is validated through a simulation example. In this example, a z-polarized resonant dipole antenna (at 6 GHz) is employed to scan the objects as shown in Figure 4.27. The figure caption shows all the dimensions and material properties. The dipole antenna scans an aperture with dimensions $r_A = 50$ mm and $z_A = 100$ mm. The number of samples are $N_\phi = 181$ and $N_z = 51$. The sampling steps along both azimuthal and longitudinal directions are well below Nyquist sampling criteria ($\lambda/4$, where λ is the smallest wavelength in this imaging example).

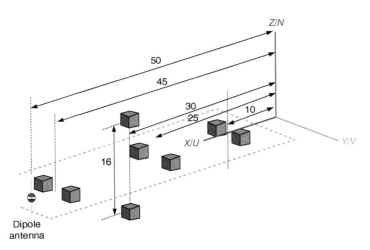

Figure 4.27 Simulation example including a resonant dipole antenna scanning over a cylindrical aperture with radius of 50 mm. Two objects are at $(r, \varphi, z) = (45$ mm, $\pm 4°, 0)$, two objects are at $(r, \varphi, z) = (25$ mm, $\pm 8°, 0)$, two objects are at $(r, \varphi, z) = (10$ mm, $\pm 15°, 0)$, and two objects are at $(r, \varphi, z) = (30$ mm, $0°, \pm 8$mm). The properties of the background medium are $\varepsilon_r = 16$ and $\sigma = 0.1$ S/m. The properties of the objects are $\varepsilon_r = 32$ and $\sigma = 0.2$ S/m. The numbers are in mm.

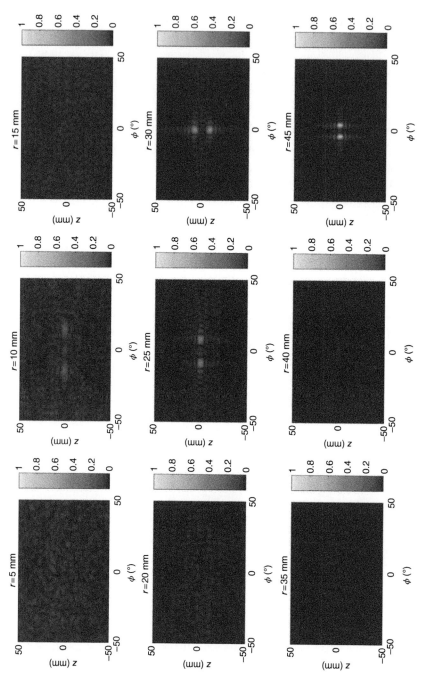

Figure 4.28 Reconstructed images for the simulation example in Figure 1.2 for nine cylindrical imaged surfaces ($N_r = 9$) from $r_1 = 5$ mm to $r_9 = 45$ mm with radial distance between these surfaces being 5 mm.

At each sampling position, backscattered field is sampled at 15 frequencies within the range of 3–10 GHz (sampling step of 0.5 GHz). According to [11], this leads to a range resolution of approximately 5 mm. Thus, the image reconstruction is performed over nine cylindrical surfaces ($N_r = 9$) from $r_1 = 5$ mm to $r_9 = 45$ mm with radial distance between these surfaces being 5 mm. Figure 4.28 shows the reconstructed images. The objects at $r_5 = 25$ mm, $r_6 = 30$ mm, and $r_9 = 45$ mm are reconstructed very well at their true positions. The objects at $r_2 = 10$ mm are reconstructed correctly but the image quality is not very good.

4.6 Three-Dimensional Holographic Imaging Using Single-Frequency Microwave Data

In Section 4.4, in order to find the contrast function $f(x, y, z_n)$ over all reconstruction planes, measurements are performed at multiple frequencies ω_l ($l = 1, 2, \dots, N_\omega$). That enables forming the systems of equations in (4.82) which then can be solved to find the contrast function. It has been shown in [134] that using N_r receivers that sample the scattered field $E_{n_r}^{sc}(x', y')$, $n_r = 1$, $\dots$, N_r, due to the object at spatially diverse positions at each position of the transmitter can also provide sufficient information for performing 3D imaging. Figure 4.29 illustrates this configuration. Similar to using multiple frequency

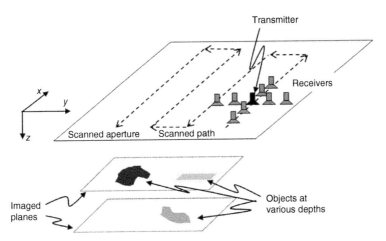

Figure 4.29 Illustration of the single-frequency 3D holographic imaging setup which contains a transmitting antenna that illuminates the inspected medium with microwave power at a single frequency. The transmitting antenna moves together with an array of N_r receiving antennas. This transmitter–receivers set scans a rectangular aperture [134].

data, using multiple receivers allows for constructing a system of equations at each spatial-frequency pair $\kappa = (k_x, k_y)$ as follows:

$$
\begin{bmatrix} \tilde{E}_1^{sc}(\kappa) \\ \vdots \\ \tilde{E}_{N_r}^{sc}(\kappa) \end{bmatrix} \approx \begin{bmatrix} \tilde{E}_1^{sc,co}(\kappa,z_1)\Delta z_1 & \cdots & \tilde{E}_1^{sc,co}(\kappa,z_{N_z})\Delta z_{N_z} \\ \vdots & \vdots & \vdots \\ \tilde{E}_{Nr}^{sc,co}(\kappa,z_1)\Delta z_1 & \cdots & \tilde{E}_{N_r}^{sc,co}(\kappa,z_{N_z})\Delta z_{N_z} \end{bmatrix} \times \begin{bmatrix} \tilde{F}(\kappa,z_1) \\ \vdots \\ \tilde{F}(\kappa,z_{N_z}) \end{bmatrix}
$$

(4.99)

where $E_{n_r}^{sc,co}(\kappa,z_n)$, $n_r = 1, ..., N_r$, is the 2D FT of the measured or simulated PSF recorded by the receiver n_r and when the CO is placed at the imaged plane $z = z_n$. The array of receivers can be on the same side of the inspected medium as the transmitter antenna or on the opposite side. The solution of this system of equations is similar to what we discussed for the wideband imaging system.

4.7 Microwave Holographic Imaging Using the Antenna Phaseless Radiation Pattern

In this section, we extend the near-field holographic techniques discussed in Section 4.2 to the case when the object is in the far zone of the antenna but the measurements are performed in the near zone of the scattering object. This algorithm employs the phaseless radiation pattern of the antenna [135]. The distinguishing features of this technique include the following. First, in the techniques discussed in Chapter 3, due to the definition of k_z as $k_z = \sqrt{4k^2 - k_x^2 - k_y^2}$, the spatial frequencies k_x and k_y cannot exceed $2k$ as this ensures real values for k_z (considering propagating waves only). However, in the holographic techniques discussed in this chapter, there is no upper limit for k_x and k_y. Larger k_x and k_y values correspond to evanescent waves that contain information about the finer details of the object. Thus, by developing an inversion technique similar to the one discussed in Section 4.2, better cross-range resolution is achieved compared to the techniques discussed in Chapter 3. Second, including the pattern of the antennas in the techniques proposed in Chapter 3 is not feasible. Here, we show that including the radiation pattern of the antenna in the processing improves the quality of the reconstructed images significantly. Also, in Section 4.2, the uncertainties in the accuracy of the simulation results for the incident field and Green's function may lead to errors in the image-reconstruction results. Here, having the antenna's phaseless radiation pattern from measurements improves the quality of the images. The phaseless radiation pattern of the antennas can be measured by well-established and cost-effective techniques (compared to vector instruments). This also provides image reconstruction on arbitrary planes without the need for a large database for the incident field and Green's function as required in the techniques discussed in Section 4.2. The concept is also useful in preventing the probe-positioning errors, which are critical in measuring the phase [135].

4.7.1 Using Phaseless Antenna Pattern in Holographic Reconstruction

The microwave holography setup considered here employs planar raster scanning. It consists of an antenna and an object in the antenna's far zone as shown in Figure 4.30. The formulation of the scattered field when using the linear Born approximation is as presented in (4.21).

As shown in Figure 4.30, an x-polarized antenna performs a 2D scan of the backscattered wave from the object on a plane at $z = 0$. From the reciprocity principle, when the transmitting antenna is also used for reception, Green's function is the same as the incident field; see [109]. Thus, the x-component of the scattered field E_x^{sc} acquired by the antenna is written as

$$E_x^{sc}(\mathbf{r}') = \iiint_V \mathbf{E}^{inc}(\mathbf{r}) \cdot \mathbf{E}^{inc}(\mathbf{r}) \left[k_s^2(\mathbf{r}) - k_b^2\right] d\mathbf{r}. \qquad (4.100)$$

When the object is in the far zone of the antenna, $\mathbf{E}^{inc}(\mathbf{r}) \cdot \mathbf{E}^{inc}(\mathbf{r})$ in (4.100) is estimated from its far-field radiation pattern. The phaseless radiation pattern of the antenna $P(\phi, \theta, \omega_l)$ when the antenna is at $(0, 0, 0)$ is assumed to be known at any measurement frequency ω_l ($l = 1, 2, \dots, N_\omega$) and at any angular position given by ϕ and θ. Then, the incident field at any point (x, y, z) in the inspected volume is written as

$$\mathbf{E}^{inc}(x,y,z;0,0,0;\omega_l) \cdot \mathbf{E}^{inc}(x,y,z;0,0,0;\omega_l) \approx 2\eta P(\phi,\theta,\omega_l)e^{-2jkr} \qquad (4.101)$$

where (r, θ, ϕ) is the representation of (x, y, z) in a spherical coordinate system.

Let the signal $E_x^{sc}(x',y',\omega_l)$ be the scattered wave received by the x-polarized dipole at $(x', y', 0)$. The incident field for the case where the antenna is at (x', y') can be obtained from that in (4.101) by a simple translation:

$$\mathbf{E}^{inc}(x,y,z;x',y',0;\omega_l) = \mathbf{E}^{inc}(x-x',y-y',z;0,0,0;\omega_l). \qquad (4.102)$$

Figure 4.30 Illustration of the 3D microwave holography setup [135].

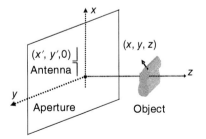

Then, the received scattered field is written as

$$E_x^{sc}(x',y',\omega_l) = \iint\int f(x,y,z)a(x'-x,y'-y,z,\omega_l)dxdydz \qquad (4.103)$$

where

$$f(x,y,z,\omega_l) = k_s^2(x,y,z,\omega_l) - k_b^2(\omega_l) \qquad (4.104)$$

$$a(x,y,z,\omega_l) = \mathbf{E}^{inc}(-x,-y,z;0,0,0;\omega_l) \cdot \mathbf{E}^{inc}(-x,-y,z;0,0,0;\omega_l). \qquad (4.105)$$

Again, we refer to $f(x, y, z, \omega_l)$ as the contrast function and we assume that it is independent of frequency (nondispersive media). In (4.103), the integral over x and y is interpreted as a 2D convolution integral. In the Fourier domain, it is written as

$$\tilde{E}_x^{sc}(k_x,k_y,\omega_l) = \int_z \tilde{F}(k_x,k_y,z)\tilde{A}(k_x,k_y,z,\omega_l)dz \qquad (4.106)$$

where $\tilde{E}_x^{sc}(k_x,k_y,\omega_l)$, $\tilde{F}(k_x,k_y,z)$, and $\tilde{A}(k_x,k_y,z,\omega_l)$ are the 2D FTs of $E_x^{sc}(x',y',\omega_l)$, $f(x, y, z)$, and $a(x, y, z, \omega_l)$, respectively; and k_x and k_y are the Fourier variables with respect to x and y, respectively.

To reconstruct the contrast function, we approximate the integral in (4.106) by a discrete sum for the N_z imaged planes:

$$\tilde{E}_x^{sc}(k_x,k_y,\omega_l) = \sum_{n=1}^{N_z} \tilde{F}(k_x,k_y,z_n)\tilde{A}(k_x,k_y,z_n,\omega_l)\Delta z \qquad (4.107)$$

where Δz is the distance between two neighboring planes.

For the setup shown in Figure 4.30, the acquired scattered wave at each scanning position is obtained by measuring the complex reflection coefficient at the antenna terminals. Thus, by performing measurements at N_ω frequencies, (4.107) can be written for all of them which leads to N_ω equations at each spatial-frequency pair $\kappa = (k_x, k_y)$ as

$$\begin{cases} \tilde{E}_x^{sc}(\kappa,\omega_1) = \tilde{A}(\kappa,z_1,\omega_1)\tilde{F}(\kappa,z_1)\Delta z + \cdots + \tilde{A}(\kappa,z_{N_z},\omega_1)\tilde{F}(\kappa,z_{N_z})\Delta z \\ \qquad\qquad\qquad\qquad \vdots \\ \tilde{E}_x^{sc}(\kappa,\omega_{N_\omega}) = \tilde{A}(\kappa,z_1,\omega_{N_\omega})\tilde{F}(\kappa,z_1)\Delta z + \cdots + \tilde{A}(\kappa,z_{N_z},\omega_{N_\omega})\tilde{F}(\kappa,z_{N_z})\Delta z. \end{cases}$$
$$(4.108)$$

The constructed system of equations is solved in the least-square sense to find $\tilde{F}(\kappa, z_n)$, $n = 1, 2, \ldots, N_z$, at each spatial frequency pair (k_x, k_y). Then, inverse 2D FT is applied to $\tilde{F}(\kappa, z_n)$, to reconstruct a 2D slice of the function $f(x, y, z_n)$ at each $z = z_n$ plane.

4.7.2 Image Reconstruction Results

In a simulation example, an x-polarized $\lambda/2$ (at 6.5 GHz) dipole antenna is employed to scan an X-shape object as shown in Figure 4.31. The object has two similar arms with length of 80 mm, width of 10 mm, and height of 1 mm. It is centered at the range position of 100 mm. The background medium is vacuum and the properties of the object are $\varepsilon_r = 2$ and $\sigma = 0.5$ S/m. The antenna scans an aperture of 300 mm × 300 mm with a sampling step of 7.5 mm at the $z = 0$ plane. The aperture should be large enough to ensure that the backscattered wave from the object is sampled until it decreases sufficiently. In practice, the scattered signal must decrease below the level of the system noise (numerical or experimental noise). In general, a larger aperture ensures that larger k_x and k_y values are acquired in the measurement leading to better cross-range resolution according to Section 3.3.2.2.

The reflection S-parameter of the antenna is acquired at each sampling position from 3 to 11 GHz with steps of 0.5 GHz and is calibrated. The phaseless radiation pattern of the antenna is obtained via simulation.

Figure 4.32a shows the reconstructed images when applying the 3D holographic technique proposed in [5]. The object is detected at $z = 120$ mm instead of $z = 100$ mm, and the cross-range resolution is not sufficient to reconstruct its shape. The poor cross-range resolution is due to the fact that k_x and k_y cannot exceed $2k$ in order to ensure real values for k_z.

Figure 4.32b shows the reconstructed images when applying the 3D holographic technique proposed in [86] where both the incident field and Green's

Figure 4.31 Simulation setup for imaging an X-shape object. Dimensions are in mm [135].

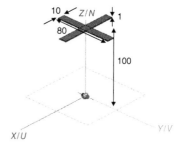

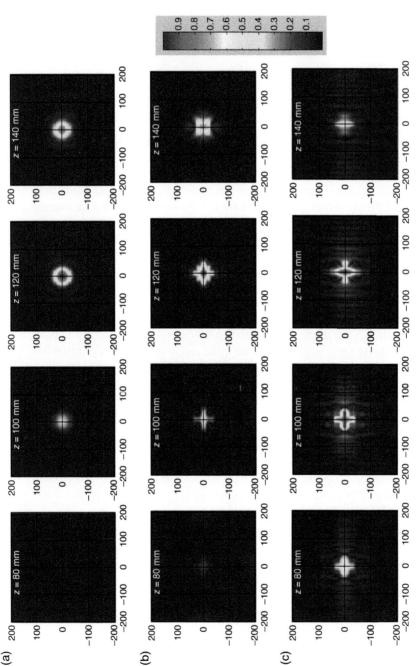

Figure 4.32 Reconstructed images for the example shown in Figure 4.31 when employing: (a) the holography technique proposed in [5], (b) the technique proposed in [86] with an approximation of the incident field and Green's function by exp(–*jkr*), and (c) the holography technique when using phaseless radiation pattern of the antenna. The horizontal axis shows the position along x in mm and the vertical axis shows the position along y in mm [135].

Figure 4.33 Illustration of the experiment for imaging of two thin rectangular copper sheets scanned by a Ka-band open-ended waveguide [135].

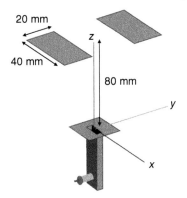

function are assumed to be e^{-jkr}. The shape of the object is reconstructed better (compared to Figure 4.32a) due to the processing of larger k_x and k_y values, allowing for the inclusion of evanescent waves. However, the focusing along the range direction is not satisfactory and the object's image at $z = 120$ mm is stronger than that at $z = 100$ mm.

Figure 4.32c shows the reconstructed images when applying the 3D holographic technique proposed here and when using the phaseless radiation pattern of the antenna. The shape of the object is reconstructed very well at $z = 100$ mm. The focusing along the range direction is also improved compared with the images in Figure 4.32a and b.

In an experiment, a Ka-band open-ended rectangular waveguide is employed to scan two rectangular copper sheets with sizes of 40 mm × 20 mm, center-to-center distance of 50 mm, and centered at range position of $z = 80$ mm, as illustrated in Figure 4.33. The background medium is air. The waveguide scans an aperture of 160 mm × 200 mm with a sampling step of 5 mm in the x and y directions at the $z = 0$ plane.

The reflection S-parameter of the antenna is acquired at each sampling position from 20 to 26 GHz with steps of 0.5 GHz and is calibrated. The holographic imaging algorithm discussed in this section is applied together with the phaseless radiation pattern of the open-ended waveguide.

When applying the 3D holographic technique proposed in [5] the objects cannot be reconstructed at all. For brevity, we do not show those images. Figure 4.34a shows the reconstructed images when applying the 3D holographic technique proposed in [86] where both the incident field and Green's function are assumed to be e^{-jkr}. It is observed that the two objects can be reconstructed at $z = 80$ mm although there is also an artifact between them. The focusing along the range direction is poor.

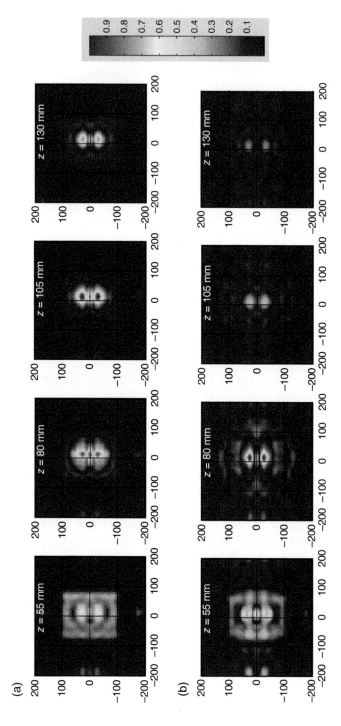

Figure 4.34 Reconstructed images for the example in Figure 4.33 when employing: (a) the technique proposed in [86] with an approximation of the incident field and Green's function by $\exp(-jkr)$, and (b) the holography technique discussed in this section when using the phaseless radiation pattern of the antenna. The horizontal axis shows the position along x in mm and the vertical axis shows the position along y in mm [135].

Figure 4.34b shows the reconstructed images when applying the 3D holographic technique discussed in this section. The two objects are reconstructed well at $z = 80$ mm. The image quality is improved significantly along both the range and the cross-range directions compared to the images obtained in Figure 4.34a.

5

Increasing the Resolution and Accuracy of Microwave/Millimeter-Wave Holography

The resolution of imaging techniques that are based on the linear Born approximation is restricted by the well-known "diffraction limit" (refer to Appendix for a brief review of diffraction limit). Most of the proposed techniques to overcome the diffraction limit exploit the evanescent field components in the extreme proximity of the imaged object. Overcoming the diffraction limit for longer imaging distances is very challenging due to the fast attenuation of the evanescent waves. Here, some techniques to overcome this limit are discussed such as the use of superoscillatory filters (SOFs) and the use of resonant scatterers in the near-field of the imaged objects. The chapter then discusses the techniques in providing quantitative holographic images of the inspected medium.

5.1 Imaging Beyond the Diffraction Limit by Applying a SOF

In this section, we first describe an approach to design a SOF and then we discuss the applications of this filter to overcome the diffraction limit of the resolution in holographic imaging [136, 137].

5.1.1 Design of 1D and 2D SOFs

SOF filter is a bandwidth-limited function such that it operates over the spatial bandwidth available through the propagating waves only, i.e. $-k_x^m \leq k_x, k_y \leq k_x^m$. However, in the spatial domain, it has a narrower width of the main lobe compared to the diffraction-limited "sinc" function. Thus, when it is applied to the images resulting from the holographic algorithm, it introduces improvement in the resolution as described later.

To design a filter with such properties, the superoscillatory (SO) waveform design concepts can be employed. Superoscillation is the phenomenon in which

Real-Time Three-Dimensional Imaging of Dielectric Bodies Using Microwave/Millimeter-Wave Holography, First Edition. Reza K. Amineh, Natalia K. Nikolova, and Maryam Ravan.
© 2019 by The Institute of Electrical and Electronics Engineers, Inc.
Published 2019 by John Wiley & Sons, Inc.

a waveform possesses variations faster than its highest spectral component. Superdirective antenna design concepts can be employed to construct a SOF. In [136, 137], the SO waveform is synthesized by the proper expansion of Tschebyscheff polynomials. This works by finding the proper sets of zeros ω_{zn} which yield the SO waveform with the desired narrow width of the main lobe and prescribed sidelobe levels.

Here, we first summarize the process of designing a SOF with $2N + 1$ spectral lines distributed evenly in the spectral domain within the bandwidth corresponding to the propagating waves. In the spatial-frequency domain, the SOF is written as

$$\tilde{F}(k_x) = \sum_{n=-N}^{N} a_n \delta\left(k_x - n\Delta k - k_x^m\right) \tag{5.1}$$

where Δk is a uniform line spacing and a_n, $n = -N, \dots, N$, are the coefficients (values of spectral lines) to be determined. The spatial variation of this filter can then be written as

$$f(x) = \prod_{n=-N}^{N} (\omega - \omega_{zn}) = \sum_{n=-N}^{N} a_n \omega^n \text{ where } \omega = e^{-jx\Delta k} \tag{5.2}$$

Defining $t = \cos(x\Delta k)$, it is easy to show that for $2N + 1$ spectral lines, with symmetrical strength around the center line, Eq. (5.2) can be written as

$$|f(x)| = a_0 + a_1 t + a_2\left(2t^2 - 1\right) + \dots + a_n T_n(t) + \dots + a_N T_N(t) \tag{5.3}$$

where $T_n(t)$ is the nth order Tschebyscheff polynomial. It is feasible to design the SOF employing the useful properties of Tschebyscheff polynomials including: (i) all zeros of $T_n(t)$ occur between $t = \pm 1$ and (ii) the maximum and minimum values of $T_n(t)$ lying in the interval of $t \in [-1, +1]$ are $T_n(t) = \pm 1$. In order to determine the coefficients $a_0, \dots, a_N$, the desired portion of the Tschebyscheff polynomial of the correct degree is shifted into the range defined by $t = \cos(H_x \Delta k)$ for the given array, where $2H_x$ is the designed spatial extent of the filter. Then, the coefficients of Eq. (5.3) are equated to those of the transformed Tschebyscheff function. Knowing the positions of the nulls, the spatial variation of the SOF is obtained from Eq. (5.3).

Furthermore, to construct a two-dimensional (2D) SOF, the simplest method is to take the array of spectral lines along the k_x-axis as an element of another similar array along the k_y-axis. Assuming similar parameters for the filter along both the k_x and k_y directions, the 2D variation of the SOF is obtained from

$$\hat{F}(k_x, k_y) = \sum_{n=-N}^{N} \sum_{m=-N}^{N} a_m a_n \delta\left(k_x - n\Delta k - k_x^m, k_y - m\Delta k - k_x^m\right) \tag{5.4}$$

5.1.2 Application of the SOF to Overcome the Diffraction-Limited Resolution

To overcome the diffraction-limited resolution, the SOF designed in the previous section can be applied to the complex-valued images obtained from the holographic imaging techniques. As described above, the SOF is designed using the range of k_x and k_y values corresponding to the propagating waves only, i.e. $-k_x^m \leq k_x \leq k_x^m$. Applying such a filter to the reconstructed values obtained from holographic imaging provides improved resolution as described later in this section. Since the filter is designed to act on the transverse spatial spectrum corresponding to the propagating waves only, the improvement in the resolution can be obtained far beyond the evanescent region.

Figure 5.1 illustrates the block diagram of the resolution enhancement algorithm in the spectral domain. For example, for a point-source, from Eq. (A.4), the estimated Fourier transform (FT) of the image obtained from holographic reconstruction is ideally 1 within the spectrum of the propagating waves. When the SOF is applied to this image, the FT of the outcome $\tilde{J}^{SOF}(k_x)$ is the product of the FT of the estimated image $\tilde{J}(k_x)$ and the FT of the filter $\tilde{F}(k_x)$. So, for a point-source, ideally we have

Figure 5.1 Block diagram of applying a SOF in the spectral domain to improve the resolution in the holographic imaging [136].

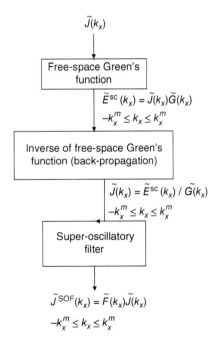

$$\tilde{J}^{SOF}(k_x) = \tilde{F}(k_x) \quad -k_x^m \le k_x \le k_x^m. \tag{5.5}$$

Thus, when taking the inverse FT of Eq. (5.5), we obtain an image $\tilde{J}^{SOF}(x)$ which has SO variation with arbitrarily narrower width (corresponding to the sharpness of the main lobe in the designed SOF) compared to the diffraction-limited "sinc" function obtained in Eq. (A.5). Similarly, for multiple point-sources, the application of the SOF leads to obtaining better resolving capability. The only limitation is that the field of view (FOV) has to be limited to avoid the appearance of the huge sidelobes that always exist in SO variations.

5.1.3 Sample Image Reconstruction Results

Here, we show a simulation example in which two z-directed dipole sources centered at $x = \pm 0.35\lambda$ in free space are imaged for $L = 10\lambda$ and $y_0 = 5\lambda$. A filter is designed with seven spectral lines in the spectral frequency range of $-0.9k \le k_x \le 0.9k$. To do that, the properties of the Tschebyscheff polynomials are employed, such that the variation in the space domain shows the narrowest central peak width and uniform sidelobes at 5% of the central peak field. Figure 5.2a shows the spectral variation of the designed SOF with seven spectral lines. Figure 5.2b shows the spatial variation of the designed filter in the interval of $-\lambda \le x \le \lambda$. This figure also shows the variation of the diffraction-limited "sinc" function within the same interval. The SOF has been designed to have a narrower width and lower sidelobe levels. Figure 5.2c shows the results of applying the designed SOF in the imaging of the two sources mentioned above. It is observed that the two sources cannot be resolved with the conventional holographic process while they can be resolved well when applying the SOF. For 2D and three-dimensional (3D) imaging results the reader can refer to [137].

5.2 Use of Resonant Scatterers in the Proximity of the Imaged Objects

In this section, we discuss a technique to perform subdiffraction imaging by using scatterers in the proximity of the imaged objects. As shown in the one-dimensional (1D) imaging configuration in Figure 4.30, we assume that the antenna performs the scan on a line placed at the far-field of the object. Thus, the spectrum of the measured waves over the scanned line corresponds to the propagating waves only. As discussed in Section 5.1, this leads to the well-known diffraction limit in the resolution of the reconstructed image along the x axis.

Figure 5.2 (a) Spatial spectrum of the designed SOF. (b) Spatial variation of the designed SOF. (c) Reconstruction of two point-sources at $x = -0.35\lambda$ and $x = 0.35\lambda$ when $L = 10\lambda$ and $y_0 = 5\lambda$ [136].

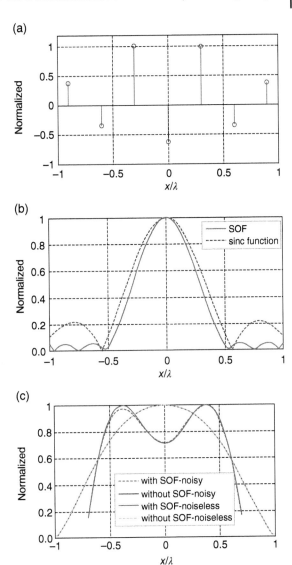

Here, we discuss a technique proposed in [138, 139] in which a resonant scatterer moves with the antenna but in the near zone of the objects (as shown in Figure 5.3). Thus, strong evanescent components generate currents on the scatterer. These currents, in turn, produce secondary scattered components, part of which reaches the antenna in the form of propagating waves. Thus, a portion of the information related to the evanescent spectrum for the primary scattered waves can be measured by the antenna leading to resolution enhancement.

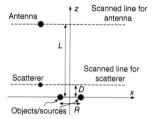

Figure 5.3 Far-field sub-diffraction imaging setup using near-field scatterers that move together with an antenna scanning in the far-field ($L \geq \lambda$) [139].

Here, first, the effect of the distance D in Figure 5.3 on the resolution is demonstrated. The scatterer is a half-wavelength wire along the y axis. The imaged objects are two PEC spheres with diameter of $\lambda/40$ and center-to-center distance of $R = 0.18\lambda$. Figure 5.4a shows the reconstructed images without using the scatterer as well as when using the scatterer and with D decreasing from $\lambda/10$ to $\lambda/30$. Significant improvement in the resolution is observed with decreasing D. Figure 5.4b shows the results when the scatterer is a quarter-wavelength in length. It is observed that this nonresonant scatterer does not improve the resolution as much as the resonant one in Figure 5.4a.

Next, the effect of having two scatterers aligned along the z axis is demonstrated. Figure 5.5a shows the images for this configuration compared with the cases when using a single scatterer or no scatterer. It is observed that when using two scatterers along the z axis, the resolution is improved significantly compared to the use of one scatterer. Further improvement in the resolution in this example indicates that a portion of the evanescent spectrum for the secondary scattered wave due to the scatterer is also converted to propagating spectrum which is measurable by the antenna.

As the last example, an imaging configuration is presented in which three scatterers are aligned along the x axis and they move together with the antenna. Figure 5.5b shows the reconstructed images with these scatterers as well as for the cases that only one scatterer or no scatterer is used. It is observed that, in contrast to the previous example, adding a third scatterer along the x axis does not improve the resolution compared to the case of two scatterers.

5.3 Quantitative Reconstruction Based on Microwave Holography

In [140, 141], a technique has been proposed to perform quantitative microwave holographic reconstruction in quasi-real time. Both the real and the imaginary parts of the permittivity distribution of a dielectric object are recovered quantitatively. The method relies on a calibration measurement of a known electrically small object. Following the approach discussed in Section 4.4 (holographic

(a)

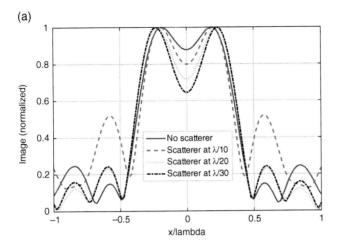

(b)

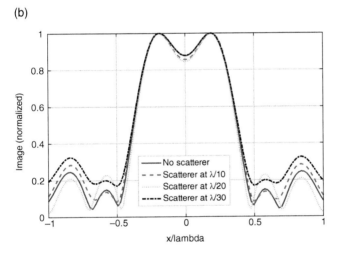

Figure 5.4 Images when changing *D* in Figure 5.3 with (a) resonant scatterer, (b) nonresonant scatterer [139].

imaging with PSF data), it is assumed that an object under test (OUT) can be viewed as a weighted combination of known point scatterers. Then, ignoring multiple scattering effects, its responses can be represented as the weighted superposition of the respective responses of the point scatterers it is built of. Since a CO is a representation of a point scatterer in the background, an OUT response can be approximated as the weighted superposition of the respective CO responses.

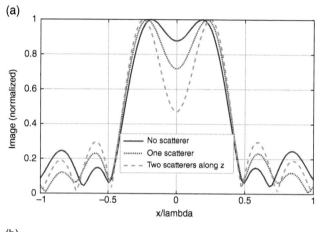

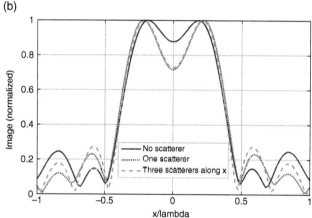

Figure 5.5 Effect of using multiple scatterers: (a) two scatterers along z axis and (b) three scatterers along x axis [139].

To generate 3D images, CO data are collected for all desired depths z_n, $n = 1$, ..., N_z, where z_n denotes the depth coordinate of the small scatterer. Also, 3D imaging requires wide-bandwidth measurements. The CO and OUT data are calibrated and Fourier transformed as described in Section 4.4. Then, the systems of equations similar to (4.82) are formed and solved at each spatial-frequency pair $\kappa = (k_x, k_y)$. Once $\tilde{F}(\kappa, z)$ is found at each point κ in Fourier space at every range location z_n, $n = 1, ..., N_z$, an inverse 2D FT recovers $f(x, y, z)$ at each z_n, i.e. slice by slice.

The linearized model of scattering, on which holography is based, allows for expressing the difference $\Delta \varepsilon_{OUT} = \varepsilon_{OUT} - \varepsilon_b$ between the complex permittivity of the OUT (ε_{OUT}) and that of the background (ε_b) as

$$\Delta\varepsilon_{OUT}(\mathbf{\kappa},z) = \delta\varepsilon_{CO}\cdot\tilde{F}(\mathbf{\kappa},z) \tag{5.6}$$

where $\delta\varepsilon_{CO} = \varepsilon_{CO} - \varepsilon_b$ is the permittivity difference of the known point scatterer in the CO. Due to the linearity of the above relationship, it also holds in real space, leading to

$$\varepsilon_{OUT}(x,y,z_n) = \delta\varepsilon_{CO}\cdot f_{OUT}(x,y,z_n) + \varepsilon_b. \tag{5.7}$$

The local accuracy of the quantitative results obtained with (5.7) is subject to a sufficiently small difference between $\varepsilon_{OUT}(x, y, z_n)$ and ε_{CO}.

As an example, simulation of an F-shaped object is generated with a background permittivity $\varepsilon_b = 1 + 0j$, an object permittivity of $\varepsilon_{OUT} = 1.2 + 0j$ and a calibration object permittivity of $\varepsilon_{CO} = 1.1 + 0j$. These data are collected in the frequency range from 3 to 16 GHz. The object is positioned at the center between the transmitting (Tx) and receiving (Rx) antennas, which are 10 cm apart. Figure 5.6 depicts the reconstruction of the object's permittivity. The accuracy of the quantitative result is very good. However, since no constraints are applied to the permittivity values, at some points nonphysical values are observed, e.g. positive imaginary part.

As another example, experimental data are collected on an OUT consisting of four scatterers that are placed at varying distances from one another in a trapezoidal configuration; see the locations indicated by the red squares in Figure 5.7. The shortest distance between the two scatterers is $\lambda/4$, where λ is the wavelength at the central frequency. The relative permittivity of all four scatterers is equal to that of the small scattering probe in the CO measurement, i.e. $\varepsilon_{CO} = 50 + 0j$, whereas the permittivity of the background medium (Styrofoam) is $\varepsilon_b \approx 1.1 + 0j$. This example tests the ability of the

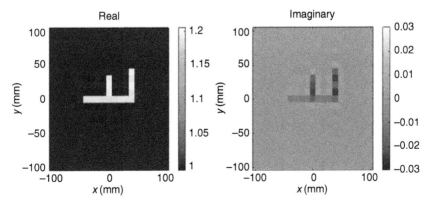

Figure 5.6 Reconstructed permittivity of an F shaped object (simulation example). Actual object has relative permittivitiy $\varepsilon_{OUT} = 1.2 + 0j$ [141].

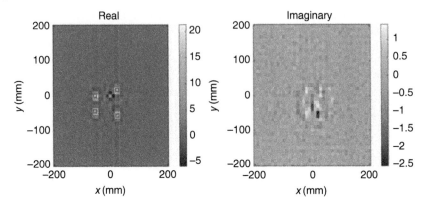

Figure 5.7 Reconstructed relative permittivities of four small scattering objects with $\varepsilon_{OUT} = 50 + 0j$ in a background of $\varepsilon_b \approx 1.1 + 0j$. The experimental setup has the two left scatterers positioned $\lambda/4$ from one another, and the two right scatterers postioned at $\lambda/2$ from one another, where λ is the wavelength at the central frequency [141].

method to handle larger permittivities (in violation of the assumed linear model of scattering) and larger noise levels.

Utilizing two waveguides (one Rx, one Tx), a 20×20 cm area is scanned across a bandwidth from 8 to 15 GHz. Note that the magnitude of the nonphysical values has increased due to the significant difference between the background and OUT permittivities, which violates the linearized scattering model. Still, the method was able to reconstruct an image of high quality.

5.4 Modifications on Holographic Imaging Improving Stability and Range Resolution

In [142], a 3D microwave holographic imaging algorithm has been proposed based on the measurement of reflection coefficient. The algorithm offers several modifications for the near-field 3D holographic imaging originally proposed in [86, 109], including: (i) an auxiliary equation is derived and exploited to effectively improve the numerical stability and the image quality, (ii) to accurately locate the unknown object in the range direction, a phase compensation method that requires only phase correction to the associated entries of the kernel matrix was proposed based on a simple ray model, (iii) considering the finite size of the scanning aperture and beamwidth of the probe antenna, a numerical low-pass filter in the spatial-frequency domain is utilized to effectively locate the worth-solving area. The limitation of the method is that it is applicable to loss-free objects.

5.4.1 Forward Model in Terms of the Open-Circuit Voltage at the Terminals of Probe Antenna

The open-circuit voltage at the port of the measuring antenna V_{oc} can be derived through the reciprocity theorem as [53, 143]

$$V_{oc}(\omega) = -\frac{1}{I_{in}(\omega)} \int_V E^{inc}(r,\omega) \cdot J(r,\omega) dr \tag{5.8}$$

where I_{in} is the input current of the antenna when it operates in a transmitting mode, E^{inc} is the incident field radiated by the antenna, and J is the induced current source inside volume V when the antenna operates in receiving mode. The open-circuit voltage of the antenna is related to both the incident field E^{inc} as the antenna operates in transmitting mode and the current source J when the antenna operates in a receiving mode. Since the open-circuit voltage incorporates the receiving properties of the probe antenna while J incorporates the properties of the transmitting antenna, the holographic imaging algorithm can be derived using (5.8) as a forward model of scattering. In [143], the forward model of scattering is derived in the form of both open-circuit voltage and scattering parameters (S-parameters), the latter being the most common response measured at microwave frequencies. We note that there is a one-to-one transformation between the antenna open-circuit voltage and the S-parameters associated with a response measured at that antenna.

The position of the scanning probe is denoted by $r' = (x', y', 0)$ and it is assumed that the probe radiates time-harmonic field of frequency ω toward the object. The position inside the object is denoted by $r = (x, y, z)$. Then, the field scattered by the object is collected by the same antenna. In [142], it is assumed that the object is made of linear, isotropic, nondispersive, and nonmagnetic material with unknown complex permittivity distribution $\varepsilon(r)$. By the volume equivalence principle, the unknown object could be replaced with the equivalent (or induced) current source J_{eq} as

$$J_{eq}(r',r,\omega) = j\omega[\varepsilon(r) - \varepsilon_b]E^{tot}(r',r,\omega) \tag{5.9}$$

where E^{tot} is the total field inside the object and ε_b is the permittivity of the background medium (assumed uniform). The open-circuit voltage of the probe antenna at r' can then be rewritten as

$$V_{oc}(r',\omega) = -\frac{j\omega}{I_{in}(\omega)} \int_V [\varepsilon(r) - \varepsilon_b]E^{inc}(r-r',\omega) \cdot E^{tot}(r',r,\omega) dr \tag{5.10}$$

where the incident field $E^{inc}(r - r', \omega)$ is the field at position r radiated by the probe at the center of the acquisition plane. The response of interest here is the open-circuit voltage $V_{oc}(r', \omega)$ but it also could be the reflection S-parameter of the probe. The latter can be calculated from the open-circuit voltage provided the input impedance of the probe is known [53]. Equation (5.10) is a nonlinear integral equation in the unknown permittivity distribution because the total field inside the object is an implicit function of the permittivity distribution.

To linearize the nonlinear problem, the first-order Born approximation is applied along with the assumption for a uniform background medium, leading to

$$E^{\text{tot}}(r',r,\omega) \approx E^{\text{inc}}(r-r',\omega). \tag{5.11}$$

By combining (5.9)–(5.11), the open-circuit voltage is written as

$$V_{\text{oc}}(r',\omega) \approx -\frac{j\omega}{I_{\text{in}}(\omega)} \int_V [\varepsilon(r)-\varepsilon_0] E^{\text{inc}}(r-r',\omega) \cdot E^{\text{inc}}(r-r',\omega) dr. \tag{5.12}$$

The forward model in terms of the open-circuit voltage in (5.12) is similar to that for the scattered field at the coaxial connector of an antenna, which was first derived in [109] and which is repeated below with modified notations for easy comparison:

$$E^{\text{sc}}(r',\omega) \approx \int_V f(r) E^{\text{inc}}(r-r',\omega) \cdot E^{\text{inc}}(r-r',\omega) dr \tag{5.13}$$

where

$$f(r) = \omega^2 \mu_0 [\varepsilon(r) - \varepsilon_0]. \tag{5.14}$$

The resolvent kernel in the integrals in (5.12) and (5.13) is similar and it is in the form of the dot product of the incident field of the antenna. This incident field represents the Green's function in the background. It also represents the total field due to the transmission into the object when employing Born's approximation. The general forward model in terms of the S-parameters, first derived in [143] and described in detail in [53], also shows that the resolvent kernel is a dot product of two field distributions: (i) the incident field $E_{\text{Rx}}^{\text{inc}}(r'_i, r, \omega)$ of the receiving antenna at r'_i if it were to act as a transmitting antenna and (ii) the total field inside the object $E_{\text{Tx}}^{\text{tot}}(r'_k, r, \omega)$ due to the transmitting antenna at r'_k:

$$S_{ik}^{\text{sc}}(r'_i, r'_k, \omega) = \frac{-j\omega}{2a_i a_k} \int_V [\varepsilon(r)-\varepsilon_b] E_{\text{Rx}}^{\text{inc}}(r'_i, r, \omega) \cdot E_{\text{Tx}}^{\text{tot}}(r'_k, r, \omega) dr \tag{5.15}$$

where a_i is the root-power wave [144][1] exciting the receiving antenna if it were to operate in a transmitting mode and a_k is the root-power wave exciting the kth port where the transmitting antenna is connected. It is clear that in the case of a reflection-coefficient measurement, where the transmission and reception are performed by the same antenna (the case considered in [142]), (5.15) reduces to

---

1 The root-power wave equals the square root of the total power carried by a traveling wave incident upon a device, e.g. an antenna. For example, a root-power wave of $a_i = 1$ $W^{1/2}$ corresponds to a one-*watt* excitation of the ith port by the transmitter connected to it.

$$S_{ii}^{sc}(r',\omega) = \frac{j\omega}{2a_i^2}\int\limits_V [\varepsilon(r) - \varepsilon_b]E_{Tx}^{inc}(r_i',r,\omega)\cdot E_{Tx}^{tot}(r_k',r,\omega)dr \qquad (5.16)$$

where $r_i' \equiv r_k' \equiv r'$. Further, if Born's approximation is applied together with the assumption of a uniform medium, (5.16) simplifies to

$$S_{ii}^{sc}(r',\omega) = \frac{j\omega}{2a_i^2}\int\limits_V [\varepsilon(r) - \varepsilon_b]E^{inc}(r' - r,\omega)\cdot E^{inc}(r' - r,\omega)dr \qquad (5.17)$$

which differs from (5.12) and (5.13) only in the (known) constants in front of the integral. In [53, 143], it is shown that the forward model of scattering in terms of any (complex-valued) scalar response, e.g. an S-parameter, an open-circuit voltage, a short-circuit current, inevitably contains the resolvent kernel formed by the dot product of the total field due to the transmitting antenna and the incident field of the receiving antenna when this antenna operates in a transmitting mode.

Expressing the resolvent kernel in terms of the scattering function g as

$$g(x,y,z;\omega) \equiv E^{inc}(-x,-y,z,\omega)\cdot E^{inc}(-x,-y,z,\omega) \qquad (5.18)$$

and using the process described in Chapter 4, 2D FT can be applied to both sides of (5.12) to obtain

$$\tilde{T}(k_x,k_y,\omega) \approx -\frac{j\omega}{I_{in}(f)}\int \tilde{F}(k_x,k_y,z)\tilde{G}(k_x,k_y,z,\omega)dz \qquad (5.19)$$

where $\tilde{T}$, $\tilde{F}$, and $\tilde{G}$ are the 2D FT of V_{oc}, f, and g, respectively. Then, by discretizing (5.19) with respect to the range direction z (with discretization step Δz) we obtain

$$\tilde{T}(k_x,k_y,\omega) \cong \sum_{n=1}^{N_z}\tilde{F}(k_x,k_y,z_n)\tilde{G}'(k_x,k_y,z_n,\omega) \qquad (5.20)$$

where

$$\tilde{G}'(k_x,k_y,z_n,\omega) \equiv -\frac{j\omega\Delta z}{I_{in}(\omega)}\tilde{G}(k_x,k_y,z_n,\omega). \qquad (5.21)$$

Writing (5.20) at N_ω frequencies, a system of equation is constructed at each spatial-frequency pair $\kappa = (k_x,k_y)$ as

$$
\begin{bmatrix}
\tilde{T}(\kappa,\omega_1) \\
\tilde{T}(\kappa,\omega_2) \\
\vdots \\
\tilde{T}(\kappa,\omega_{N_\omega})
\end{bmatrix}
=
\begin{bmatrix}
\tilde{G}'(\kappa,z_1,\omega_1) & \tilde{G}'(\kappa,z_2,\omega_1) & \cdots & \tilde{G}'(\kappa,z_{N_z},\omega_1) \\
\tilde{G}'(\kappa,z_1,\omega_2) & \tilde{G}'(\kappa,z_2,\omega_2) & \cdots & \tilde{G}'(\kappa,z_{N_z},\omega_2) \\
\vdots & \vdots & \ddots & \vdots \\
\tilde{G}'(\kappa,z_1,\omega_{N_\omega}) & \tilde{G}'(\kappa,z_2,\omega_{N_\omega}) & & \tilde{G}'(\kappa,z_{N_z},\omega_{N_\omega})
\end{bmatrix}
\begin{bmatrix}
\tilde{F}(\kappa,z_1) \\
\tilde{F}(\kappa,z_2) \\
\vdots \\
\tilde{F}(\kappa,z_{N_z})
\end{bmatrix}.
$$
$$(5.22)$$

The solution process is the same as the one described in Chapter 4.

5.4.2 Applying an Auxiliary Equation for Numerical Stability

It has been proposed in [142] that if the background material and the object are lossless, then an auxiliary equation can be employed to increase the numerical stability of the holographic imaging problem. In this case, the contrast function $f(x,y,z)$ is real-valued, and the corresponding 2D FT $\tilde{F}(\kappa,z)$ is a Hermitian function, i.e.

$$\tilde{F}\left(-k_x, -k_y, z_n\right) = \overline{\tilde{F}\left(k_x, k_y, z_n\right)} \tag{5.23}$$

where $\overline{\tilde{F}\left(k_x, k_y, z_n\right)}$ is the complex conjugate of $\tilde{F}\left(k_x, k_y, z_n\right)$. Considering (5.20), we have

$$\tilde{T}\left(-k_x, -k_y, \omega\right) \cong \sum_{n=1}^{N_z} \tilde{F}\left(-k_x, -k_y, z_n\right) \tilde{G}'\left(-k_x, -k_y, z_n, \omega\right). \tag{5.24}$$

Taking the complex conjugate of both sides of (5.24) and substituting (5.23) into it leads to

$$\overline{\tilde{T}\left(-k_x, -k_y, \omega\right)} \cong \sum_{n=1}^{N_z} \tilde{F}\left(k_x, k_y, z_n\right) \overline{\tilde{G}'\left(-k_x, -k_y, z_n, \omega\right)} \tag{5.25}$$

Equation (5.25) can be used as an auxiliary equation for solving the holographic imaging problem. In other words, besides (4.81), (5.25) provides additional equations for solving for the unknown contrast function. This increases the numerical stability of the problem. Although the auxiliary equation in (5.25) is subject to the lossless assumption, it has been described in [142] that this auxiliary equation can still help in imaging of lossy objects with dielectric loss tangent up to approximately 0.1.

5.4.3 Phase Compensation Method

In [142], a method called the phase compensation method has been proposed to improve the accuracy of localization of the object along the range direction. This has been performed via employing the different round-trip phases. In order to implement this method, first note that for various z_n ($n = 1, \dots, N_z$), $G'(k_x, k_y, z_n)$ corresponds to the case in which the waves propagate toward different positions on the z-axis, namely $z_1, z_2, \dots,$ and z_{Nz}, and then return to their starting points. In [142], it is assumed that the major difference between these cases is simply the round-trip phase, $\exp(-j2k_z z_n)$. Thus, to implement the phase compensation method, a simple relation is established between the functions G' on different observation z-planes. Assuming that only the propagation components are measurable, G' is confined within the $2k$-circle in the 2D spatial-frequency domain and G' at z_n is related to G' at z_1 as

$$\tilde{G}'\left(k_x,k_y,z_n,\omega\right) \cong \tilde{G}'\left(k_x,k_y,z_1,\omega\right)e^{-jk_z'(z_n-z_1)} \tag{5.26}$$

where

$$k_z' \equiv \sqrt{(2k)^2 - k_x^2 - k_y^2}. \tag{5.27}$$

Here, it is assuming that the antenna is directive and consequently kx and ky values are small. It is also known that a function in the form of $g(x) = (1-x^2)^{1/2}$ changes slowly as long as x is small. These lead to approximating k_z' as a constant value in (5.26). The phase term in (5.26) is then considered as the round-trip phase delay due to the free-space propagation of a plane wave. When there is an object on the path of the waves, the presence of this dielectric material increases the phase delay. This additional phase delay could be roughly compensated based on a simple ray model. First, consider a small object with refractive index n and thickness Δz centered at the z_j plane as shown in Figure 5.8. For G' on the z_j plane, the ray penetrates into the object and is reflected back from the center of the object at $z = z_j$. The additional phase delay ϕ_m can be derived from (5.26) as

$$\phi_m = k_{z,m}'(n-1)\Delta z/2 \tag{5.28}$$

where the subscript m indicates the mth frequency. This additional phase delay ϕ_m can be employed to modify the G' associated with the z_j plane for all frequencies. Thus, the coefficient matrix in (5.22) can be written as

$$\begin{bmatrix} \cdots & \tilde{G}'\left(\kappa,z_{j-1},\omega_1\right) & \tilde{G}'\left(\kappa,z_j,\omega_1\right)e^{-j\phi_1} & \tilde{G}'\left(\kappa,z_{j+1},\omega_1\right) & \cdots \\ \cdots & \tilde{G}'\left(\kappa,z_{j-1},\omega_2\right) & \tilde{G}'\left(\kappa,z_j,\omega_2\right)e^{-j\phi_2} & \tilde{G}'\left(\kappa,z_{j+1},\omega_2\right) & \cdots \\ \cdots & \vdots & \vdots & \vdots & \cdots \\ \cdots & \tilde{G}'\left(\kappa,z_{j-1},\omega_{N_\omega}\right) & \tilde{G}'\left(\kappa,z_j,\omega_{N_\omega}\right)e^{-j\phi_{N_\omega}} & \tilde{G}'\left(\kappa,z_{j+1},\omega_{N_\omega}\right) & \cdots \end{bmatrix}.$$

$$\tag{5.29}$$

Figure 5.8 Illustration of the ray model for phase compensation [142].

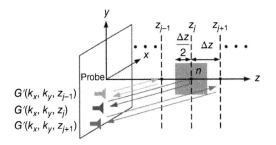

By using the coefficient matrix (5.29) in (5.22), a phase-compensated least-square problem is obtained. The auxiliary equation described in Section 5.4.2 can be still used to increase the numerical stability of the least-square problem. Also, note that two rounds of solving the least-square problems are needed. In the first round (without phase compensation), the approximate values of the refractive index n and range position z_j of the unknown object are estimated. Based on these approximate values, the second round can then be executed with phase compensation. In has been shown in [142] that using the phase-compensation method improves the accuracy of localizing the object along the range direction and, in general, leads to better quality images. This also helps to have a better quantitative estimation of the permittivity values.

5.4.4 Numerical Low-Pass Filter in Spatial-Frequency Domain

In the holographic imaging, a least-square problem is solved at each (k_x, k_y) pair. It is important to find the values of (k_x, k_y) pair for which the corresponding least-square problem is worth solving. For example, when the measurements are implemented in the far field of the antenna, the measureable values of (k_x, k_y) are inside the circle of radius $2k_{max}$, where k_{max} is the maximum wavenumber corresponding to the highest frequency. Outside this circle, G' vanishes and the resultant coefficient matrix in (5.22) is nearly a zero matrix, making the associated least-square problem meaningless. Thus, to select only the useful values of (k_x, k_y) pair for which the corresponding least-square problems are worth solving, a numerical low-pass filter in the spatial-frequency domain is used.

In general, the (k_x, k_y) pairs for the worth-solving least-square problems are determined following the criterion that the probe antenna, when scanning to the position that is the most distant from the center of the scanning aperture, could still "see" the object [5, 53]. To simplify the discussion, let us consider the center frequency ω_{center}, the wavenumber of which is denoted as k_c, along with the center observation plane z_{center}. Since the radiation pattern of the probe antenna and the dimensions of the scanning aperture are two key factors in designing the numerical low-pass filter in the spatial-frequency domain, they will be discussed separately.

Figure 5.9a depicts both the x–z and y–z plane patterns of the probe antenna when it scans to the outermost positions on the x-axis and the y-axis and its main beam can still cover the object located at $(0,0,z_{center})$. The largest viewing angles or beamwidths for the two planar patterns are $\theta_{1,x}$ and $\theta_{1,y}$, respectively, and the associated wave vectors pointing from the probe antenna to the object are $(k_c\sin(\theta_{1,x}/2),0,k_z)$ and $(0,-k_c\sin(\theta_{1,y}/2),k_z)$, respectively. The corresponding angular spectra can then be acquired by plane-wave expansion, and hence $G'(-2k_c\sin(\theta_{1,x}/2),0,z_{center})$ and $G'(0,2k_c\sin(\theta_{1,y}/2),z_{center})$ are low-pass filtered

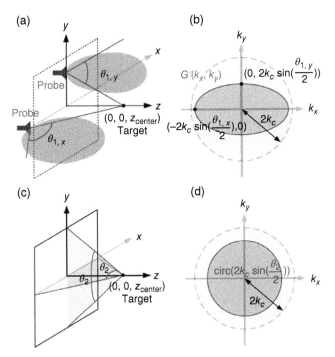

Figure 5.9 (a) Illustration of the beamwidths of the probe antenna and (b) the associated low-pass filtering in the spatial-frequency domain. (c) Illustration of the finite extent of the scanning aperture and (d) the associated low-pass filtering in the spatial-frequency domain [142].

as shown in Figure 5.9b. Besides, according to (5.26), we have $|G'(z_1)| \cong |G'(z_2)| \cong \dots \cong |G'(z_{Nz})|$ and thus can choose any G' for designing the numerical filter.

Second, given the dimensions of the square scanning aperture, the angle θ_2 subtended by the aperture from the point $(0,0,z_{center})$, as shown in Figure 5.9c, can be obtained readily. The corresponding wave vectors pointing from the mid-points of the four sides of the aperture boundary to the object at $(0,0,z_{center})$ are $(\pm k_c \sin(\theta_2/2),0,k_z)$ and $(0, \pm k_c \sin(\theta_2/2),k_z)$. The associated worth-solving area in the spatial-frequency domain, as shown in Figure 5.9d, can be roughly estimated as a circle function, $\text{circ}(2k_c \sin(\theta_2/2))$, defined as

$$\text{circ}(\rho) = \begin{cases} 1, k_x^2 + k_y^2 \le \rho^2 \\ 0, k_x^2 + k_y^2 \le \rho^2. \end{cases} \tag{5.30}$$

The intersection of the elliptical area depicted in Figure 5.9b and the circular area in Figure 5.9d defines the worth-solving area in the spatial-frequency domain. In this way, the range of the desired (k_x, k_y) pair can be narrowed down, and the number of least-square problems to be solved can be reduced drastically. This not only saves the computational time but also eliminates noise from outside the worth-solving area.

5.4.5 Simulation Results

In the examples shown in Figure 5.10a and b, two different probe antennas designed at 10 GHz, namely a half-wavelength dipole and a patch antenna, are used to reconstruct images of a pair of identical dielectric cuboids, the dimensions of which are 18 mm × 18 mm × 3 mm and the perittivity of which is $\varepsilon_r = 2$. The center frequency of the system is chosen at 10 GHz. Frequency bandwidth is set to be 20% (9–11 GHz), within which the radiation pattern of the patch antenna remains well shaped but is slightly distorted near 11 GHz. The patch antenna is designed on a 31-mil RT/Duroid 5880 dielectric slab and the detailed parameters are given in Figure 5.10b. The size of the scanning aperture is 360 mm × 360 mm, which is large enough to reduce the truncation error due to the omni-directional pattern of the dipole. The spatial sampling steps $\Delta x = \Delta y = 6$ mm satisfy the sampling criteria of at least two samples per (shortest) wavelength.

The spacing between the scanning aperture and the objects is set at 60 mm such that the object lies in the radiating near-field region of the probe antenna. Also, we set $\Delta z = \lambda_{center}/10 = 3$ mm, $N_z = 3$ for an acceptable condition number of 10.6.

The image slices on the three chosen observation planes, $z = 57, 60$, and 63 mm, reconstructed via the setups in Figure 5.10a and b are shown in Figure 5.11a and b, respectively. Clearly, the images retrieved via the two setups resemble each other

(a)   (b)

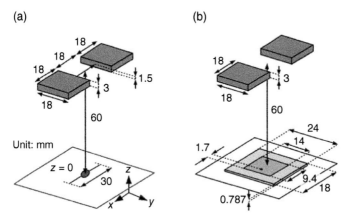

Figure 5.10 Scanning setups for a pair of dielectric cuboids using a half-wavelength dipole [142].

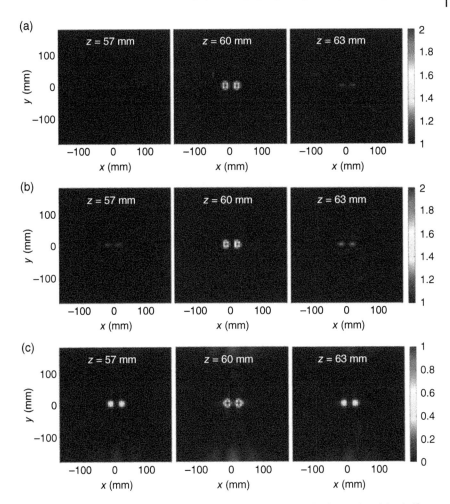

Figure 5.11 Reconstructed images using the algorithm in [142] when using: (a) a half-wavelength dipole (at 10 GHz) and (b) a patch antenna as the probe. (c) Reconstructed images using a half-wavelength dipole as the probe and the reflection-coefficient version of the algorithm proposed in [109]. (Observation planes: $z = 57$, 60, and 63 mm.)

and the object indeed appears in the image slice of the $z = 60$ mm plane with $\varepsilon_r = 2$ in both cases. Figure 5.11c shows the images reconstructed via the setup in Figure 5.10a but using the single-antenna version of the algorithm in [109]. The image quality is not as good as in Figure 5.11a and b, especially on the $z = 57$ and 63 mm planes.

6

Conclusion

In this short book, we reviewed the major developments in microwave and millimeter-wave holographic imaging. Currently, holographic techniques are considered to be one of the most practical and successful microwave and millimeter-wave imaging (MMI) techniques due to their widespread use in security screening. The recent developments have paved the way toward using these techniques in extreme near-field imaging applications such as tissue imaging, nondestructive testing, and even in microscopy. Since the inversion process in these techniques mainly involves the use of the Fourier transform (FT) and the solution of small systems of equations, these techniques yield results in real-time or quasi real-time. This feature, makes them attractive for applications such as motion tracking and functional tissue imaging.

The imaging time for holographic MMI systems can be divided into two parts: data acquisition time and image-reconstruction time. As mentioned above, the latter can be short due to the use of fast Fourier transforms (FFTs) and/or the solution of small systems of equations. This process can be expedited even further if parallel processing is employed, because the small systems of equations are solved independently for each spatial frequency pair (k_x, k_y). However, data acquisition consumes much longer time than the image-reconstruction time. This is due to the need for scanning over an aperture in order to form a synthetic aperture. There have been already attempts to reduce the data acquisition time in these techniques. For example, sequentially switched linear arrays have been used to expedite the measurements. Electronic switching along one array axis followed by mechanical scanning along the orthogonal axis allows for dense sampling of a two-dimensional aperture within seconds even when the whole human body is scanned. This approach has been employed for several applications including concealed weapon detection, ground penetrating radar, and nondestructive inspection and evaluation.

Another solution toward reducing the data acquisition time has been demonstrated in [145] where a sparse multi-static array technology has been

Real-Time Three-Dimensional Imaging of Dielectric Bodies Using Microwave/Millimeter-Wave Holography, First Edition. Reza K. Amineh, Natalia K. Nikolova, and Maryam Ravan.
© 2019 by The Institute of Electrical and Electronics Engineers, Inc.
Published 2019 by John Wiley & Sons, Inc.

developed that reduces the number of antennas required to sample one axis of the aperture. This allows a significant reduction in cost and complexity of the linear-array-based imaging system. The sparse array has been specifically designed to be compatible with FT-based image reconstruction techniques.

Yet, another solution has been initiated in [146] which proposes a new generation of active arrays for microwave tissue imaging. Each element of the array consists of a sensor that integrates a printed slot antenna with a chip low-noise amplifier (LNA) and a mixer, with its output signal being at an intermediate frequency (IF) of 30 MHz. The measured dynamic range of the bias-switched radio sensor is 109 and 118 dB with IF resolution bandwidths of 1 kHz and 100 Hz, respectively. Further miniaturization is forthcoming by integrating the chip LNA with the slot antenna on a common printed circuit board (PCB) and integrating the wireless-transmission module supplying the local-oscillator (LO) signal.

As a final note, the progress in expediting the data acquisition in holographic MMI techniques will transform them to a totally new level where data acquisition over an entire aperture can be performed within a second. This, in turn, leads to true real-time imaging that provides fast qualitative or quantitative images of the inspected objects. This will pave the way toward the use of these techniques in applications such as functional tissue imaging, object tracking, volumetric multiphase fluid monitoring, etc. In general, we believe that this is a leading technology among all MMI techniques in terms of practicality and speed. Its future applications will grow in accordance with the rapidly increasing demand for sensing, detection, and imaging devices.

Appendix

Diffraction Limit for the Spatial Resolution in Far-Field Imaging

In holographic imaging, the coherent data are collected on a two-dimensional (2D) aperture plane and holographic algorithms are employed to reconstruct images of the sources or objects on the imaged plane/planes. In general, single-frequency data are employed to perform 2D imaging at a single-range position whereas the acquisition of wideband data allows for three-dimensional (3D) imaging of the sources or objects at various range positions.

Here, for simplicity we start our formulation assuming a linear aperture and a linear imaged domain (both along the x direction) as shown in Figure A.1. We also assume single-frequency data acquisition for now which allows for one-dimensional (1D) imaging over the inspected region.

With reference to Figure A.1, for sources on the imaged line $y = 0$, the measured scattered field $E^{sc}(x)$ on the scanned line $-L \leq x \leq L$ is obtained from

$$E^{sc}(x) = \int_{x'=-L}^{x'=+L} J(x')G(x-x')dx' \tag{A.1}$$

where J is the current density of the sources on the imaged line and G is the Green's function. Equation (A.1) is a convolution integral. Taking the Fourier transform (FT) of both sides of Eq. (A.1) gives

$$\tilde{E}^{sc}(k_x) = \tilde{J}(k_x)\tilde{G}(k_x) \tag{A.2}$$

where $\tilde{E}^{sc}$, $\tilde{J}$, and $\tilde{G}$ are the FTs of E^{sc}, J, and G, respectively.

In the 1D holographic imaging algorithm, knowing $\tilde{E}^{sc}$ and $\tilde{G}$ in Eq. (A.2), the estimated FT of the image of the source $\tilde{J}$ is obtained as

$$\tilde{J}(k_x) = \tilde{E}^{sc}(k_x)/\tilde{G}(k_x). \tag{A.3}$$

The reconstructed image $J(x)$ is then obtained by taking the inverse FT of $\tilde{J}(k_x)$.

Since the scanned line has a finite size, the maximum measurable wave number in a far-field imaging system (when measurement of the scattered fields

Real-Time Three-Dimensional Imaging of Dielectric Bodies Using Microwave/Millimeter-Wave Holography, First Edition. Reza K. Amineh, Natalia K. Nikolova, and Maryam Ravan.
© 2019 by The Institute of Electrical and Electronics Engineers, Inc.
Published 2019 by John Wiley & Sons, Inc.

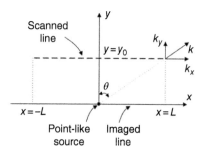

Figure A.1 Simplified illustration of the setup for 1D holographic imaging [136].

are performed in the far-field of the object) corresponding to the x axis is $k_x^m = k \sin \theta$ as illustrated in Figure A.1, where θ is the angle subtended by the aperture (assuming that the beamwidth of the antenna is larger, otherwise θ will be the beamwidth of the antenna). Thus, Eq. (A.3) is valid only within the spectral range of $-k_x^m \leq k_x \leq k_x^m$.

To study the resolution of this approach, we follow the reconstruction process for a point-source on the x-axis, i.e. $J(x) = \delta(x)$ and $\tilde{J}(k_x) = 1$. Following the above expressions, it is easy to show that the estimated FT of the image of this source $\tilde{J}(k_x)$ is obtained as

$$\tilde{J}(k_x) = 1 \;\; \text{for} - k_x^m \leq k_x \leq k_x^m \tag{A.4}$$

Therefore, the reconstructed image is a "sinc" function:

$$\tilde{J}(x) = 2k_x^m \text{sinc}\left(\frac{k_x^m}{\pi} x \right). \tag{A.5}$$

The first null of the *sinc* function occurs at $x_0 = \lambda/(2 \sin \theta)$, where λ is the operation wavelength. The diffraction limited resolution is the minimum distance between two point sources so that they can be resolved well in the reconstructed image. It is approximately $\delta x \approx 2x_0$. In practice, when the scanned aperture is sufficiently large, $\delta x \approx \lambda$. When imaging objects, the maximum measurable wave number is doubled, i.e. $-2k_x^m \leq k_x \leq 2k_x^m$ due to the fact that the path between the antenna and the object is travelled twice. This leads to a spatial resolution which is one-half of that for imaging sources, i.e. $\delta x \approx x_0$. In some literature, the diffraction limit is defined as the distance between the major axis of the *sinc* function and its first null. Thus, the diffraction limit for resolving two sources and two objects is computed as $\delta x \approx x_0$ and $\delta x \approx x_0/2$, respectively.

The holographic approach for 2D imaging is very similar to the 1D imaging described above. Thus, the diffraction limit in the resolution is defined similarly. For 3D holographic imaging, however, wideband data has to be acquired and the processing involves more complicated steps and approximations. A detailed derivation of the resolution limits, the diffraction limit included, for the 3D imaging of both sources and scatterers is available in [53].

References

1 Food and Drug Administration (FDA) Medical X-ray Imaging. https://www.fda.gov/radiation-emitting-products/medical-imaging/medical-x-ray-imaging

2 Lopez-Sanchez, J.M. and Fortuny-Guasch, J. (May 2000). 3-D radar imaging using range migration techniques. *IEEE Trans. Antennas Propag.* 48 (5): 728–737.

3 Moreira, A., Prats-Iraola, P., Younis, M. et al. (Mar. 2013). A tutorial on synthetic aperture radar. *IEEE Geosci. Remote Sens. Mag.* 1 (1): 6–43.

4 Jol, H.M. (2008). *Ground Penetrating Radar Theory and Applications.* Elsevier.

5 Sheen, D.M., McMakin, D.L., and Hall, T.E. (2001). Three-dimensional millimeter-wave imaging for concealed weapon detection. *IEEE Trans. Microw. Theory Tech.* 49 (9): 1581–1592.

6 Zoughi, R. (2000). *Microwave Non-Destructive Testing and Evaluation.* Dordrecht, The Netherlands: Kluwer.

7 Amin, M.G. (ed.) (2016). *Through-the-Wall Radar Imaging.* CRC Press.

8 Nikolova, N.K. (Dec. 2011). Microwave imaging for breast cancer. *IEEE Microw. Mag.* 12 (7): 78–94.

9 Nozokido, T., Bae, J., and Mizuno, K. (Mar. 2001). Scanning near-field millimeter-wave microscopy using a metal slit as a scanning probe. *IEEE Trans. Microw. Theory Tech.* 49 (3): 491–499.

10 Appleby, R. and Anderton, R.N. (2007). Millimeter-wave and submillimeter-wave imaging for security and surveillance. *Proc. IEEE* 95 (8): 1683–1690.

11 Sheen, D.M., McMakin, D., and Hall, T.E. (2010). Near-field three-dimensional radar imaging techniques and applications. *Appl. Opt.* 49 (19): E83–E93.

12 Ahmed, S.S., Genghammer, A., Schiessl, A., and Schmidt, L.-P. (2013). Fully electronic E-band personnel imager of 2 m^2 aperture based on a multistatic architecture. *IEEE Trans. Microw. Theory Tech.* 61 (1): 651–657.

13 Sheen, D.M. and Hall, T.E. (2014). Reconstruction techniques for sparse multistatic linear array microwave imaging. *Passive and Active Millimeter-Wave Imaging XVII*, vol. 9078 of *Proceedings of SPIE* (June 2014), p. 12.

Real-Time Three-Dimensional Imaging of Dielectric Bodies Using Microwave/Millimeter-Wave Holography, First Edition. Reza K. Amineh, Natalia K. Nikolova, and Maryam Ravan.
© 2019 by The Institute of Electrical and Electronics Engineers, Inc.
Published 2019 by John Wiley & Sons, Inc.

14 Imaging Technology, Transportation Security Administration (TSA). https://web.archive.org/web/20100106043039/http://www.tsa.gov/approach/tech/imaging_technology.shtm

15 Laskey, M. (2010). An assessment of checkpoint security: are our airports keeping passengers safe? In: *House Homeland Security Subcommittee on Transportation Security & Infrastructure Protection*, House Hearing, 111 Congress, Second Session, U.S. Government Publishing Office.

16 Harwood, M. (2014). Companies seek full-body scans that ease health, privacy concerns. *Security Management*.

17 Appleby, R. (2012). Passive millimeter–wave imaging and how it differs from terahertz imaging. *Philos. Trans. R. Soc. A Math. Phys. Eng. Sci.* 362 (1815): 379–393.

18 Nikolova, N.K. (2014). Microwave biomedical imaging. In: *Wiley Encyclopedia of Electrical and Electronics Engineering* (ed. K. Chang), 1–22. Wiley.

19 Semenov, S.Y. and Corfield, D.R. (2008). Microwave tomography for brain imaging: feasibility assessment for stroke detection. *Int. J. Antennas Propag.* 2008: 254830-1–254830-8.

20 Mobashssher, A.T., Mahmoud, A., and Abbosh, A.M. (2016). Portable wideband microwave imaging system for intracranial hemorrhage detection using improved back-projection algorithm with model of effective head permittivity. *Sci. Rep.* 6 (20459): 1–16.

21 Persson, M., Fhager, A., Trefna, H.D. et al. (Nov. 2014). Microwave-based stroke diagnosis making global prehospital thrombolytic treatment possible. *IEEE Trans. Biomed. Eng.* 61 (11): 2806–2817.

22 Lin, J.C. and Clarke, M.J. (May 1982). Microwave imaging of cerebral edema. *Proc. IEEE* 70 (5): 523–524.

23 Henriksson, T., Klemm, M., Gibbins, D. et al. (2011). Clinical trials of a multi-static UWB radar for breast imaging. *Proc. Loughborough Antennas Propag. Conf.*, Loughborough (November 2011), pp. 1–4.

24 Fear, E.C., Bourqui, J., Curtis, C. et al. (May 2013). Microwave breast imaging with a monostatic radar-based system: a study of application to patients. *IEEE Trans. Microw. Theory Tech.* 61 (5): 2119–2128.

25 Hagness, S.C., Fear, E.C., and Massa, A. (2012). Guest editorial: special cluster on microwave medical imaging. *IEEE Antennas Wirel. Propag. Lett.* 11: 1592–1597.

26 Colton, D. and Monk, P. (1995). The detection and monitoring of leukemia using electromagnetic waves: numerical analysis. *Inverse Prob.* 11 (2): 329–341.

27 Meaney, P.M., Goodwin, D., Golnabi, A.H. et al. (Dec. 2012). Clinical microwave tomographic imaging of the calcaneus: a first-in-human case study of two subjects. *IEEE Trans. Biomed. Eng.* 59 (12): 3304–3313.

28 Semenov, S.Y., Svenson, R.H., Boulysev, A.E. et al. (Sep. 1996). Microwave tomography: two-dimensional system for biological imaging. *IEEE Trans. Biomed. Eng.* 43 (9): 869–877.

29 Semenov, S.Y. (Jun. 2000). Three-dimensional microwave tomography: experimental imaging of phantoms and biological objects. *Trans. Microw. Theory Tech.* 48 (6): 1071–1074.

30 Semenov, S.Y., Bulyshev, A.E., Posukh, V.G. et al. (Mar. 2003). Microwave tomography for detection/imaging of myocardial infarction. I. Excised canine hearts. *Ann. Biomed. Eng.* 31 (3): 262–270.

31 Semenov, S.Y., Posukh, V.G., Bulyshev, A.E. et al. (2006). Microwave tomographic imaging of the heart in intact swine. *J. Electromag. Waves Appl.* 20 (7): 873–890.

32 Brovoll, S., Berger, T., Paichard, Y. et al. (2013). Time-lapse imaging of human heartbeats using UWB radar. *IEEE Biomed. Circuits Syst. Conf.*, pp. 142–145.

33 Brovoll, S., Berger, T., Paichard, Y. et al. (Oct. 2014). Time-lapse imaging of human heart motion with switched array UWB radar. *IEEE Trans. Biomed. Circuits Syst.* 8 (5): 704–715.

34 Salvador, S.M., Fear, E.C., Okoniewski, M., and Matyas, J.R. (Aug. 2010). Exploring joint tissues with microwave imaging. *IEEE Trans. Microw. Theory Tech.* 58 (8): 2307–2313.

35 Mohammed, B.J., Abbosh, A.M., Mustafa, S., and Ireland, D. (Jan. 2014). Microwave system for head imaging. *IEEE Trans. Instrum. Meas.* 63 (1): 117–123.

36 EM Tensor, Vienna, Austria. (2014, Dec.). www.emtensor.com

37 Medfield Diagnostics AB, Gothenburg, Sweden. (2014, Dec.). www.medfielddiagnostics.com

38 Micrima, Bristol, U.K. (2014, Dec.). www.micrima.com

39 Lazebnik, M., McCartney, L., Popovic, D. et al. (2007). A large-scale study of the ultrawideband microwave dielectric properties of normal, benign and malignant breast tissues obtained from cancer surgeries. *Phys. Med. Biol.* 52 (20): 6093–6115.

40 Sugitani, T., Kubota, S., Kuroki, S. et al. (2014). Complex permittivities of breast tumor tissues obtained from cancer surgeries. *Appl. Phys. Lett.* 104 (25): 253702-1–253702-5.

41 Mashal, A., Sitharaman, B., Li, X. et al. (Aug. 2010). Toward carbon-nanotube-based theranostic agents for microwave detection and treatment of breast cancer: enhanced dielectric and heating response of tissue-mimicking materials. *IEEE Trans. Biomed. Eng.* 57 (8): 1831–1834.

42 Bellizzi, G., Bucci, O.M., and Catapano, I. (Sep. 2011). Microwave cancer imaging exploiting magnetic nanoparticles as contrast agent. *IEEE Trans. Biomed. Eng.* 58 (9): 2528–2536.

43 Bucci, O.M., Bellizzi, G., Catapano, I. et al. (2012). MNP enhanced microwave breast cancer imaging: measurement constraints and achievable performances. *IEEE Antennas Wirel. Propag. Lett.* 11: 1630–1633.

44 Scapaticci, R., Bellizzi, G., Catapano, I. et al. (Apr. 2014). An effective procedure for MNP-enhanced breast cancer microwave imaging. *IEEE Trans. Biomed. Eng.* 61 (4): 1071–1079.

45 Bevacqua, M.T. and Scapaticci, R. (Feb. 2016). A compressive sensing approach for 3D breast cancer microwave imaging with magnetic nanoparticles as contrast agent. *IEEE Trans. Med. Imag.* 35 (2): 665–673.

46 Bucci, O.M., Bellizzi, G., Borgia, A. et al. (Aug. 2017). Experimental framework for magnetic nanoparticles enhanced breast cancer microwave imaging. *IEEE Access* 5: 16332–16340.

47 Bahr, A.J. (1982). *Microwave Nondestructive Testing Methods*. Newark, NJ: Gordon and Breach.

48 Mazlumi, F., Sadeghi, S.H.H., and Moini, R. (Oct. 2006). Interaction of an open-ended rectangular waveguide probe with an arbitraryshape surface crack in a lossy conductor. *IEEE Trans. Microw. Theory Tech.* 54 (10): 3706–3711.

49 Ghasr, M., Carroll, B.J., Kharkovsky, S. et al. (Oct. 2006). Millimeter wave differential probe for nondestructive detection of corrosion precursor pitting. *IEEE Trans. Instrum. Meas.* 55 (5): 1620–1627.

50 Kharkovsky, S., Case, J.T., Abou-Khousa, M.A. et al. (Aug. 2006). Millimeter wave detection of localized anomalies in the space shuttle external fuel tank insulating foam. *IEEE Trans. Instrum. Meas.* 55 (4): 1250–1257.

51 Kim, S., Yoo, H., Lee, K. et al. (Apr. 2005). Distance control for near-field scanning microwave microscope in liquid using a quartz tuning fork. *Appl. Phys. Lett.* 86: 153506.

52 Kharkovsky, S. and Zoughi, R. (Apr. 2007). Microwave and millimeter wave nondestructive testing and evaluation - overview and recent advances. *IEEE Instrum. Meas. Mag.* 10 (2): 26–38.

53 Nikolova, N.K. (Jul. 2017). *Introduction to Microwave Imaging*. Cambridge University Press.

54 Rubek, T., Meaney, P.M., Meincke, P., and Paulsen, K.D. (2007). Nonlinear microwave imaging for breast-cancer screening using gauss-Newton's method and the CGLS inversion algorithm. *IEEE Trans. Antennas Propag.* 55 (8): 2320–2331.

55 Semenov, S.Y., Bulyshev, A.E., Abubakar, A. et al. (Jul. 2005). Microwave-tomographic imaging of the high dielectric-contrast objects using different image-reconstruction approaches. *IEEE Trans. Microw. Theory Tech.* 53 (7): 2284–2294.

56 Joachimowicz, N., Pichot, C., and Hugonin, J. (Dec. 1991). Inverse scattering: an iterative numerical method for electromagnetic imaging. *IEEE Trans. Antennas Propag.* 39 (12): 1742–1753.

57 Chew, W.C. and Wang, Y.M. (Jun. 1990). Reconstruction of two-dimensional permittivity distribution using the distorted born iterative method. *IEEE Trans. Med. Imaging* 9 (2): 218–225.

58 Franchois, A. and Tijhuis, A.G. (Apr. 2003). A quasi-Newton reconstruction algorithm for a complex microwave imaging scanner environment. *Radio Sci.* 38 (2): VIC 12-1–VIC 12-13.

59 Souvorov, A.E., Bulyshev, A.E., Semenov, S.Y. et al. (Nov. 1998). Microwave tomography: a two-dimensional newton iterative scheme. *IEEE Trans. Microw. Theory Tech.* 46 (11): 1654–1659.

60 Franchois, A. and Pichot, C. (Feb. 1997). Microwave imaging-complex permittivity reconstruction with a Levenberg-Marquardt method. *IEEE Trans. Antennas Propag.* 45 (2): 203–215.

61 Zaeytijd, J.D., Franchois, A., Eyraud, C., and Geffrin, J. (Nov. 2007). Full-wave three-dimensional microwave imaging with a regularized gauss-newton method-theory and experiment. *IEEE Trans. Antennas Propag.* 55 (11): 3279–3292.

62 Meaney, P.M., Paulsen, K.D., Pogue, B.W., and Miga, M.I. (Feb. 2001). Microwave image reconstruction utilizing log-magnitude and unwrapped phase to improve high-contrast object recovery. *IEEE Trans. Med. Imaging* 20 (2): 104–116.

63 Kleinman, R.E. and van den Berg, P.M. (Jan. 1992). A modified gradient method for two-dimensional problems in tomography. *J. Comput. Appl. Math.* 42: 17–35.

64 Bulyshev, A.E. (2000). Three-dimensional microwave tomography: theory and computer experiments in scalar approximation. *Inverse Prob.* 16: 863–875.

65 Shumakov, D.S., Tajik, D., Beaverstone, A.S., and Nikolova, N.K. (2017). Real-time quantitative reconstruction methods in microwave imaging, Chapter 17. In: *The World of Applied Electromagnetics – in Appreciation of Magdy Fahmy Iskander* (ed. A. Lakhtakia and C.M. Furse), 415–442. Springer.

66 Tu, S., McCombe, J.J., Shumakov, D.S., and Nikolova, N.K. (2015). Fast quantitative microwave imaging with resolvent kernel extracted from measurements. *Inverse Prob.* 31 (4): 045007.

67 Shumakov, D.S. and Nikolova, N.K. Fast quantitative microwave imaging with scattered-power maps. *IEEE Trans. Microw. Theory Tech.* PP (99): 1–11. https://doi.org/10.1109/TMTT.2017.2697383.

68 Tajik, D., Pitcher, A.D., and Nikolova, N.K. (2017). Comparative study of the Rytov and Born approximations in quantitative microwave holography. *Prog. Electromagn. Res.* 79: 1–19.

69 Li, X., Bond, E.J., Van Veen, B.D., and Hagness, S.C. (2005). An overview of ultra-wideband microwave imaging via space-time beamforming for early-stage breast-cancer detection. *IEEE Antennas Propag. Mag.* 47 (1): 19–34.

70 Tabbara, W., Duchene, B., Pichot, C. et al. (1988). Diffrcation tomography: contribution to the analysis of some applications in microwave and ultrasonics. *Inverse Prob.* 4: 305–331.

71 Zhang, Y., Tu, S., Amineh, R.K., and Nikolova, N.K. (2012). Resolution and robustness to noise of the sensitivity-based method for microwave imaging with data acquired on cylindrical surfaces. *Inverse Prob.* 28 (11): 115006.

72 Sheen, D.M., McMakin, D.L., and Hall, T.E. (2007). Near field imaging at microwave and millimeter wave frequencies. *IEEE/MTT-S International Microwave Symposium*, pp. 1693–1696.

73 Sachs, J. (2012). *Handbook of Ultra-Wideband Short-Range Sensing.* Wiley–VCH Verlag & Co. KGaA.

74 Taylor, J.D. (2016). *Ultrawideband Radar: Applications and Design.* CRC Press.

75 Kak, A.C. and Slaney, M. (1988). *Principles of Computerized Tomographic Imaging.* IEEE Press.

76 Baribaud, M., Dubois, F., Floyrac, R. et al. (1982). Tomographic image reconstitution of biological objects from coherent microwave diffraction data. *IEE Proc. H (Microwaves, Optics and Antennas)* 129 (6): 356–359.

77 Bolomey, J.C., Izadnegahdar, A., Jofre, L. et al. (Nov. 1982). Microwave diffraction tomography for biomedical applications. *IEEE Trans. Microw. Theory Tech.* 30 (11): 1998–2000.

78 Pichot, C., Jofre, L., Peronnet, G., and Bolomey, J. (1985). Active microwave imaging of inhomogeneous bodies. *IEEE Trans. Antennas Propag.* 33 (4): 416–425.

79 Song, Y. and Nikolova, N.K. (2008). Memory-efficient method for wideband self-adjoint sensitivity analysis. *IEEE Trans. Microw. Theory Tech.* 56 (8): 1917–1927.

80 Liu, L., Trehan, A., and Nikolova, N. (2010). Near-field detection at microwave frequencies based on self-adjoint response sensitivity analysis. *Inverse Prob.* 26 (10): 105001.

81 Gabor, D. (May 1948). A new microscope principle. *Nature* 161 (4098): 777–778.

82 Leith, E.N. and Upatnieks, J. (1962). Reconstructed wavefronts and communication theory. *J. Opt. Soc. Am.* 52: 1123–1130.

83 Farhat, N.H. and Guard, W.R. (Sep. 1971). Millimeter wave holographic imaging of concealed weapons. *Proc. IEEE* 59 (9): 1383–1384.

84 Farhat, N.H. (1972). Microwave holography and its applications in modern aviation. *Proc. SPIE Eng. Appl. Holography Symp.*, pp. 295–314.

85 Tricoles, G. and Farhat, N.H. (Jan. 1977). Microwave holography: applications and techniques. *Proc. IEEE* 65 (1): 108–121.

86 Amineh, R.K., Ravan, M., Khalatpour, A., and Nikolova, N.K. (2011). Three-dimensional near-field microwave holography using reflected and transmitted signals. *IEEE Trans. Antennas Propag.* 59 (12): 4777–4789.

87 Amineh, R.K., McCombe, J., Khalatpour, A., and Nikolova, N.K. (2015). Microwave holography using point-spread functions measured with calibration objects. *IEEE Trans. Instrum. Meas.* 64 (2): 403–417.

88 Dallinger, A., Schelkshorn, S., and Detlefsen, J. (Sep. 2006). Efficient ω-k-algorithm for circular SAR and cylindrical reconstruction areas. *Adv. Radio Sci.* 4: 85–91.

89 Tan, W.X., Hong, W., Wang, Y., and Wu, Y. (2011). A novel spherical wave three-dimensional imaging algorithm for microwave cylindrical scanning geometries. *Prog. Electromagn. Res.* 111: 43–70.

90 Case, J.T., Ghasr, M.T., and Zoughi, R. (May 2013). Nonuniform manual scanning for rapid microwave nondestructive evaluation imaging. *IEEE Trans. Instrum. Meas.* 62 (5): 1250–1258.

91 Gabor, D. (1949). Microscopy by reconstructed wave-fronts. *Proc. R. Soc. Lond. A Math. Phys. Sci.* 197 (1051): 454–487.

92 Born, M. and Wolf, E. (1999). *Principles of Optics*, 7e. Cambridge University Press.

93 Leith, E.N. and Upatnieks, J. (1963). Wavefront reconstruction with continuous–tone objects. *J. Opt. Soc. Am.* 53 (12): 1377–1381.

94 Hildebrand, B.P. and Brenden, B.B. (1972). *An Introduction to Acoustical Holography*. Plenum Press.

95 Williams, E.G. (1999). *Fourier Acoustics: Sound Radiation and Nearfield Acoustical Holography*. Academic Press.

96 Hayek, S.I. (2008). Nearfield acoustical holography. In: *Handbook of Signal Processing in Acoustics*, (ed. D. Havelock, S. Kuwano, and M. Vorländer) 1129–1139. Springer.

97 Dooley, R. (1965). X-band holography. *Proc. IEEE* 53 (11): 1733–1735.

98 Duffy, D. (1966). Optical reconstruction from microwave holograms. *J. Opt. Soc. Am.* 56 (6): 832–832.

99 Tricoles, G. and Rope, E.L. (1967). Reconstructions of visible images from reduced–scale replicas of microwave holograms. *J. Opt. Soc. Am.* 57 (1): 97–99.

100 Kock, W. (1968). Stationary coherent (hologram) radar and sonar. *Proc. IEEE* 56 (12): 2180–2181.

101 Kock, W.E. (1975). Microwave holography. In: *Engineering Applications of Lasers and Holography*, (ed. W. E. Kock) 179–223. Springer.

102 Leith, E.N. (Sep. 1971). Quasi-holographic techniques in the microwave region. *Proc. IEEE* 59 (9): 1305–1318.

103 Farhat, N.H. (1986). Microwave holography and coherent tomography. In: *Medical Applications of Microwave Imaging* (ed. L.E. Larsen and J.H. Jacobi), 66–81. New York: IEEE Press.

104 Smith, D., Leach, M., Elsdon, M., and Foti, S.J. (Feb. 2007). Indirect holographic techniques for determining antenna radiation characteristics and imaging aperture fields. *IEEE Antennas Propag. Mag.* 49 (1): 54–67.

105 Leach, M., Smith, D., and Skobelev, S.P. (Oct. 2008). A modified holographic technique for planar near-field antenna measurements. *IEEE Trans. Antennas Propag.* 56 (10): 3342–3345.

106 Elsdon, M., Yurduseven, O., and Smith, D. (2013). Early stage breast cancer detection using indirect microwave holography. *Prog. Electromag. Res.* 143: 405–419.

107 Yurduseven, O., Smith, D., Livingstone, B. et al. (Jul. 2013). Investigations of resolution limits for indirect microwave holographic imaging. *Int. J. RF Microwave Comput. Aided Eng.* 23 (4): 410–416.

108 Yurduseven, O. (2014). Indirect microwave holographic imaging of concealed ordnance for airport security imaging systems. *Prog. Electromagn. Res.* 146: 7–13.

109 Amineh, R.K., Khalatpour, A., Xu, H. et al. (Dec. 2012). Three-dimensional near-field microwave holography for tissue imaging. *Int. J. Biomed. Imaging* 2012: 291494.

110 Amineh, R.K., Ravan, M., McCombe, J., and Nikolova, N.K. (Jul. 2013). Three-dimensional microwave holographic imaging employing forward-scattered waves only. *Int. J. Antennas Propag.* 2013: 1–15.

111 Iizuka, K. and Gregoris, L.G. (1970). Application of microwave holography in the study of the field from a radiating source. *Appl. Phys. Lett.* 17: 509–512.

112 Kock, W.E. (1972). Real-time detection of metallic objects using liquid crystal microwave holograms. *Proc. IEEE* 60: 1104.

113 Iizuka, K. (1969). Microwave holography by photoengraving. *Proc. IEEE* 57: 812414.

114 Lohmann, A.W. (1969). Data economy in holography. *J. Opt. Soc. Am.* 59: 482.

115 Kock, W.E. and Harvey, F.K. (1951). A photographic method for displaying sound wave and microwave space patterns. *Bell Syst. Tech. J.* 20: 564–587.

116 Schejbal, V., Kovarik, V., and Cermak, D. (2008). Synthesized-reference-wave holography for determining antenna radiation characteristics. *IEEE Antennas Propag. Mag.* 50 (5): 71–83.

117 Smith, D., Yurduseven, O., Livingstone, B., and Schejbal, V. (2014). Microwave imaging using indirect holographic techniques. *IEEE Antennas Propag. Mag.* 56 (1): 104–117.

118 Joy, E. and Paris, D. (1972). Spatial sampling and filtering in near-field measurements. *Trans. Antennas Propag.* AP-20 (3): 253–261.

119 Collins, H.D., McMakin, D.L., Hall, T.E., and Gribble, R.P. (1995). Real-time holographic surveillance system. US Patent 5, 455,590, 3 October 1995.

120 Soumekh, M. (Sep. 1991). Bistatic synthetic aperture radar inversion with application in dynamic object imaging. *IEEE Trans. Signal Process.* 39: 2044–2055.

121 Soumekh, M. (Jan. 1992). A system model and inversion for synthetic aperture radar imaging. *IEEE Trans. Image Process.* 1: 64–76.

122 Soumekh, M. (1994). *Fourier Array Imaging*. Englewood Cliffs, NJ: Prentice-Hall.

123 Chew, W.C. (1990). *Waves and Fields in Inhomogeneous Media*. IEEE Press.

124 Sheen, D.M., McMakin, D.L., Hall, T.E., and Severtsen, R.H. (1999). Real-time wideband cylindrical holographic surveillance system. US Patent 5, 859,609, January 1999.

125 Detlefsen, J., Dallinger, A., Huber, S., and Schelkshorn, S. (2005). Effective reconstruction approaches to millimeter-wave imaging of humans. *Proc. of the URSI General Assembly*, New Delhi, India, (October 2005), pp. 23–29.

126 Dudley, D.G. (1994). *Mathematical Foundations for Electromagnetic Theory*. The IEEE/OUP Series on Electromagnetics Wave Theory. IEEE Press.

127 Ravan, M., Amineh, R.K., and Nikolova, N.K. (2010). Two-dimensional near-field microwave holography. *Inverse Prob.* 26: 055011.

128 Ravan, M., Amineh, R.K., and Nikolova, N.K. (2010). Microwave holography for near-field imaging. *IEEE Int. Symp. on Antennas and Propag. and USNC/URSI National Radio Science Meeting (AP-S/URSI)*.

129 Banos, A. (1966). *Dipole Radiation in the Presence of a Conducting Half-Space*. New York: Pergamon Press.

130 Amineh, R.K., Khalatpour, A., and Nikolova, N.K. (2012). Three-dimensional microwave holographic imaging using co- and cross-polarized data. *IEEE Trans. Antennas Propag.* 60 (7): 3526–3531.

131 Tikhonov, A.N. and Arsenin, V.Y. (1977). *Solutions of Ill-Posed Problems*. New York: Wiley.

132 Walterscheid, I., Brenner, A.R., and Ender, J.H.G. (Sep. 2004). Results on bistatic synthetic aperture radar. *Electron. Lett.* 40 (19): 1224–1225.

133 Amineh, R.K., Ravan, M., Trehan, A., and Nikolova, N.K. (Mar. 2011). Near-field microwave imaging based on aperture raster scanning with TEM horn antennas. *IEEE Trans. Antennas Propag.* 59 (3): 928–940.

134 Amineh, R.K., Ravan, M., Sharma, R., and Baua, S. (2018). Three-dimensional holographic imaging using single frequency microwave data. *Int. J. Antennas Propag.* 2018: 1–14.

135 Amineh, R.K., McCombe, J., and Nikolova, N.K. (2012). Microwave holographic imaging using the antenna phaseless radiation pattern. *IEEE Antennas Wirel. Propag. Lett.* 11: 1529–1532.

136 Amineh, R.K. and Eleftheriades, G.V. (2013). Imaging beyond the diffraction limit by employing a super-oscillatory filter. *IEEE Int. Symp. on Antennas and Propag. and USNC/URSI National Radio Science Meeting (AP-S/URSI)*.

137 Amineh, R.K. and Eleftheriades, G.V. (2013). 2D and 3D sub-diffraction source imaging with a superoscillatory filter. *Opt. Express* 21 (7): 8142–8157. https://doi.org/10.1364/OE.21.008142.

138 Patel, A. and Amineh, R.K. (2017). Sub-diffraction holographic imaging with resonant scatterers. *Progress In Electromagnetics Research M* 59: 1–7.

139 Patel, A. and Amineh, R.K. (2018). Sub-diffraction holographic imaging with resonant scatterers in proximity of the objects. *IEEE Int. Symp. on Antennas and Propag. and USNC/URSI National Radio Science Meeting (AP-S/URSI)*.

140 Tajik, D., Pitcher, A.D., and Nikolova, N.K. (2017). Comparative study of the Rytov and Born approximations in quantitative microwave holography. *Progress in Electromagnetic Research (PIER) B* 79: 1–19.

141 Tajik, D., Thompson, J.R., Beaverstone, A.S., and Nikolova, N.K. (2016). Real-time quantitative reconstruction based on microwave holography. *IEEE AP-S/URSI Int. Symp. on Antennas and Propagation* (June 2016).

142 Tsai, C., Chang, J., Yang, L.O., and Chen, S. (Jan. 2018). 3-D microwave holographic imaging with probe and phase compensations. *IEEE Trans. Antennas Propag.* 66 (1): 368–380.

143 Beaverstone, A.S., Shumakov, D.S., and Nikolova, N.K. (Apr. 2017). Integral equations of scattering for scalar frequency-domain responses. *IEEE Trans. Microw. Theory Tech.* 64 (4): 1120–1132.

144 Pozar, D.M. (2012). *Microwave Engineering*, 4e. Wiley.

145 Sheen, D.M. and Hall, T.E. (2014). Reconstruction techniques for sparse multistatic linear array microwave imaging. *Proc. SPIE 9078, Passive and Active Millimeter-Wave Imaging XVII, 90780I.* doi:https://doi.org/10.1117/12.2053814.

146 Foroutan, F. and Nikolova, N.K. (2018). Active sensor for microwave tissue imaging with bias-switched arrays. *Sensors* 18 (5): pii: E1447. https://doi.org/10.3390/s18051447.

Index

Real-Time Three-Dimensional Imaging of Dielectric Bodies Using Microwave/Millimeter-Wave Holography, First Edition. Reza K. Amineh, Natalia K. Nikolova, and Maryam Ravan.
© 2019 by The Institute of Electrical and Electronics Engineers, Inc.
Published 2019 by John Wiley & Sons, Inc.

IEEE Press Series on
RF and Microwave Technology

Series Editor: **George Kizer**, *Telecommunications Consultant, Dallas, Texas, USA*

Wireless technology has become a worldwide game-changer. It has redefined how we—and machines—interact in ways unimaginable a few years ago. The IEEE Press Series on RF and Microwave Technology spans the latest development in this area. While integrating established practices with emerging developments, the series encompasses devices, systems and enabling technology over the wide span of activity. This series will be of interest to students, practicing engineers and anyone interested in how this world-changing technology is evolving.

Computational Electromagnetic-Aerodynamics · Joseph J.S. Shang

Real-Time Three-Dimensional Imaging of Dielectric Bodies Using Microwave/ Millimeter-Wave Holography · Reza K. Amineh, Natalia K. Nikolova, and Maryam Ravan